THE SURVIVAL HANDBOOK

THE SURVIVAL HANDBOOK

Peter Darman

This book is dedicated to my parents

ISBN 1 56865 178 3

Printed in USA

This Edition produced for
DOUBLEDAY BOOK & MUSIC CLUBS, INC

All artworks by Graham Bingham
Design by Cooper Wilson Design

CONTENTS

BASICS

Simple principles lie at the heart of survival. These include mental attitude, what to wear and what to carry. As a survivor, you must master these basic survival skills, especially the psychology of survival, for they will ultimately determine whether you live or die.

THE PSYCHOLOGY OF SURVIVAL

To survive, you need survival skills – but skills alone will not save you. You need attitude, and it has to be the right attitude. All the knowledge in the world will count for nothing unless you also have the will to survive.

The will to survive is all-important in a survival situation. It is well known that the mind may give up before the body, but with a determination to survive individuals can give themselves an enormous head start in an emergency situation.

Think about this. No matter how bad the situation you find yourself in, remember that you have immediate resources to help you out of your predicament: your mental abilities and your physical attributes. Put them both to work effectively and you can achieve results.

HOW TO FOCUS YOUR MIND TO SURVIVE

Two of the greatest threats to survival come from your own mind. They are a desire for comfort and a passive outlook. If not countered quickly, they can result in the demoralisation and death of survivors. Fortunately, they can be dealt with quite easily by any survivor.

A desire for comfort is a consequence of modern urban living conditions. Western living standards have made people 'soft', in the sense that they are, for the most part, cushioned from threats from nature and the environment. Most Westerners - you included, probably - live and work in warm, secure buildings, have access to sophisticated health care, and have a guaranteed supply of food and water.

In an emergency scenario you will probably have none of these things, at least not initially. You may may only possess the clothes you stand up in, with no immediate food, water or shelter. The sudden disappearance of the comforts you have taken for granted is a great shock to the system and can lead to severe demoralisation. So how do you combat the mental anguish caused by the loss of those things that you regard as essential to life?

First, tell yourself that Western-style comforts are not essential to survival. Be obstinate. You really can get by without air conditioning, junk food and the TV. Second, tell yourself this: your present discomfort will be nothing compared to the extreme discomfort you will experience if you just sit down, bemoan your lot and do nothing.

A passive outlook is also a consequence of living in the West. One result of living in bureaucratic states is that the individual does not have to make life-or-death decisions. Individual decision-making is reduced to the mundane and banal. Initiative is stifled and most individuals have a passive, almost sheep-like, outlook. However, in a survival situation you could be on your own. If you are you will have to make important decisions. If this appears daunting, remember that the result of doing nothing will probably be your death. On the other hand, you can take control of the situation and live. Which do you prefer? *DO NOT LIE DOWN AND DIE, GET MOTIVATED AND ACTIVE!*

DEFEAT THE ENEMIES OF SURVIVAL

There are other enemies of survival of a more physical nature, and you must be aware of them so you can take effective counter-measures:

PAIN: pain is the body's way of telling you that something is wrong. It is uncomfortable and can weaken the will to survive. However, it becomes much more tolerable if you understand its source and nature, recognise it as something to be tolerated, and concentrate on other tasks. Remember, pain will seem worse if you do nothing but lie down and think about it.

COLD: slows down the flow of blood and makes you sleepy. It also dulls the mind. This is dangerous; you should immediately seek shelter and build a fire.

US ARMY TIPS

PERSONAL QUALITIES ESSENTIAL TO THE SUCCESSFUL OUTCOME OF A SURVIVAL SITUATION

The US Army knows from long experience what a man needs to get through a survival situation. Have you got what it takes?

- Ability to concentrate the mind.
- Ability to improvise.
- Ability to live with yourself.
- Ability to adapt to a situation.
- Ability to remain calm.
- Ability to be optimistic, while at the same time preparing for the worst.
- Ability to understand your own fears and worries and being able to deal with and overcome them.

THIRST: like hunger it can dull the mind. It is important to keep up the intake of water. If water is scarce cut down on food intake; the body uses water to carry off wastes from food.
HUNGER: can result in a loss of weight, weakness, dizziness and blackouts, slowed heart rate, increased sensitivity to the cold and increased thirst. Obviously this can be countered by food intake.
FATIGUE: can bring on lethargy, poor mental outlook, such as hopelessness, lack of a goal and boredom. can It is important for the survivor to take sufficient rest.
BOREDOM: can result in a lack of interest, feelings of strain and depression (especially with no hope of relief). To overcome boredom you must keep your goal - survival - uppermost in your mind and realise how the tasks you undertake fit into your overall survival plan.
LONELINESS: being alone can obviously bring on loneliness, which can then cause a sense of helplessness and despair. This can be overcome by keeping busy and becoming more self-sufficient.
FRUSTRATION: can be countered by channelling your energies into positive and obtainable goals. Complete the easier tasks before attempting the more challenging ones. In addition, you must accept the situation you are in and act accordingly. Do not have unrealistic goals. Do not sit down and worry - keep busy.

DON'T PANIC

If you have just survived a car or aircraft crash, or a shipping accident, your mind will be in a state of disorientation. You may be injured, there may be injured people and dead bodies all around you. Even though you may be in pain and very frightened, try to do two things:

- ☐ Get away from the immediate vicinity of the wreckage and any danger of being burned or injured by explosions or fires.
- ☐ Once you are out of danger - stay put. The worst thing you can do is to blindly stumble into unknown territory, especially if it is dark. This will only result in getting lost and risking serious injury. Instead, sit down, and do not panic. Analyse the situation you are in as calmly as possible.

IF YOU CAN STAY CALM IT WILL PAY DIVIDENDS LATER ON.

ASSESSING EMERGENCY SITUATIONS

Once you have got yourself away from the danger area, there are a number of things you must do. Take note of these points, they will benefit you enormously:

- ☐ Find a place that is sheltered so you can sit down and think.
- ☐ Evaluate your surroundings. Every environment in the world has its own rhythm and pattern. Get to know the one you are in.
- ☐ Assess your physical condition: do you have any wounds, do your require extra clothing, or food and water?

☐ Assess the equipment you may have. Have you got anything that is useful? What condition is it in?

☐ Do not act in haste, this can only result in general disorientation and the danger of losing some of your equipment.

REMEMBER, IT IS MUCH EASIER TO MAKE AN ACCURATE ASSESSMENT OF A SITUATION IN DAYLIGHT THAN AT NIGHT.

BRITISH SAS TIPS

PROCEDURE FOR THE IMMEDIATE AFTERMATH OF A CRASH

British Special Air Service advice to men on the correct procedure to follow in the immediate aftermath of an air crash:

- Do not leave the site of the crash immediately unless you are in danger.
- Treat the injured immediately.
- Separate the dead from the living.
- Salvage anything you can, such as equipment, food, clothing and water.
- Take stock of the location you find yourself in.
- Do NOT move at night unless absolutely necessary.

FORMULATING YOUR BLUEPRINT FOR SURVIVAL

You must now put together a plan for you personal survival. Good planning and preparation enable the survivor to overcome difficulties and dangers and will keep you alive.

When putting together your survival plan, remember that if your were in an aircraft crash or shipping accident, your position will probably have been reported just prior to the accident. Therefore, the rescue services will know the approximate position of any survivors, and, if they have been alerted, search teams will almost certainly be on their way. But you must still make some decisions. You cannot just sit down and wait for rescue; rely on yourself to stay alive, not on other people. For many people, making decisions is extremely difficult - however, you must take control of the situation.

Displaying calmness and having confidence in yourself will obviously inspire confidence and cooperation in others, and will make the implementation of a plan much easier - people will follow leaders..

Situation appraisal is most important when putting together a plan for survival. You must be as objective as possible and weigh up all the positive and negative aspects of the situation you are in. For example, if your are in arctic conditions the extreme cold will pose obvious dangers, but you will have an abundance of water in the shape of snow and ice. In the tropics you will have good food sources, but the heat and humidity will be hazards.

Main priorities in a survival situation are shelter, fire and water. If you have an abundance of water and fuel for a fire, ie wood, you will also have access to shelter-building material. If this is the case you will have all three of your main priorities and should stay where you are in the short term.

Once again it is important to stress the need to think of every aspect of the situation you are in. For example, if you were involved in an aircraft crash you might be able to retrieve materials from the wreckage (being careful to ensure there is no risk of fire or explosions before you venture into the aircraft's remains). The foam-rubber seats make excellent fire-starter material or warm foot protection. If the aircraft had a battery it can be used to start fires. In addition, the aircraft's tyres can be used to start an excellent signal fire (thick black smoke can be seen over long distances). By walking away from the aircraft in the immediate aftermath of the accident you would have deprived yourself of all these useful items of equipment.

To travel or to stay put can often be a difficult decision in a survival situation. That said, there are powerful reasons to stay put. First, as mentioned above, the rescue services will know of your general location and will already be looking for you. Second, by walking you will be burning a lot of calories, you will be subject to the elements (which may be harsh), you may not be able to set up a permanent, effective signalling system (your morale will plummet if you see an aircraft but have no way of making it spot you), and you risk literally walking yourself to death. However, if, for whatever reason, you decide to travel, you should formulate a plan that includes the following:

☐ The direction you will travel in.
☐ A method of keeping on the course you have determined.
☐ A schedule of how long you will walk each day.
☐ A method of signalling.

Remember, you need to allow enough time at the end of each day to establish a camp and a signalling system should an aircraft fly over.

If you decide to stay put, on the other hand, then your plan must include the following, in order of priority:

☐ The establishment of a signalling system.
☐ The location and style of your permanent camp.
☐ Determining your water source.
☐ What will constitute your diet.

It is very important to establish a signalling system first, as a spotter aircraft may overfly your position at any time and you must be ready. To this end, it is wise to build your camp near the signalling system.

Now you only need survival and medical skills and knowledge to enable you to live and eventually get back to civilisation, and these will be provided in full in the following chapters.

WHAT TO WEAR

Adventurers need clothing that will withstand the rigours of hostile weather and terrain. Here are the principles you should follow when selecting clothing for backpacking in wild areas. Above all, do not compromise when it comes to clothing.

If you have just survived a plane crash or similar accident, the chances are that you will be dressed in light, comfortable clothes that are totally unsuitable to survival situations. In this case you must improvise (see the chapter on improvising clothing, tools and weapons). However, if you are a backpacker or similar adventurer, you should be better clothed and equipped to deal with a survival situation, and there is really no excuse for wearing sub-standard clothing and carrying poor equipment.

Today, because of the explosion in outdoor leisure activities, there is a bewildering variety of clothing available to the backpacker, with a corresponding diversity in quality and price. It is impossible here to give a detailed breakdown of the range of clothing currently available. Nevertheless, some guiding principles can be provided that will enable you to make a wise choice when selecting outdoor wear.

HOW TO CHOOSE YOUR SURVIVAL CLOTHES

Above all, you must select the proper clothing for the job. For example, clothing that is suitable for a Sunday afternoon stroll in a temperate climate will not stand up to the severe conditions encountered in the arctic, tropics or desert. Do not skimp when it comes to clothing – *IT WILL BE THE MAIN FACTOR THAT PROTECTS YOU FROM THE COLD, WET AND WIND.*

But how do you know what to kit yourself out in? Simple: read this book, go through the plethora of magazines dedicated to outdoor pursuits, go to camping exhibitions, go to camping and survival shops and discuss your requirements with staff who have a knowledge about such things. In short, like all special forces units, undertake detailed research before your embark on an operation.

In this way you will not find out the hard way that the clothing you purchased is totally unsuitable for the expedition you are currently undertaking. A little prior planning will pay dividends.

Gore-tex is an excellent material for outdoor clothing. It is a 'breathable' material which allows perspiration vapour to exit but prevents water from entering. This is achieved via the microporous Gore-tex membrane, which has nine billion pores per square inch. These pores are 20,000 times smaller than a droplet of water but 700 times bigger than a molecule of water vapour. This prevents the entry of wind and water but allows perspiration to escape freely. Gore-tex clothing is not cheap, but what price do you put on your life?

The layer principle offers the maximum protection and flexibility in all types of climate. The principle is very simple: dead air is the best form of insulation, and the best way of creating it is to trap it between layers of clothing. The more layers you wear, the greater the insulating effect. Temperature control is very easy: all you do is add or take away layers according to your wants.
Remember, overheating can be as much of a problem as being cold. If you sweat when it's cold, the body chills when you stop sweating, and your sweat-soaked clothing will act as a conductor to draw away body heat into the air. It is important that you prevent this. Here are the layers you should wear:

- ☐ Next to the skin you should wear thermal underwear, so-called 'long johns'.
- ☐ Over this should be worn a woollen or wool mixture shirt.
- ☐ On top of this will be a woollen or good woven fibre sweater or jacket (woven fibre tends to be better because it is warmer and more windproof).
- ☐ Then have a fibre-filled or holofil (man-made material) jacket. Down is not recommended, when it gets wet it tends to lose its insulating properties.
- ☐ The final layer must be windproof and waterproof.

US AIR FORCE TIPS

GENERAL CARE OF CLOTHING – THE 'COLDER' PRINCIPLE

USAF pilots learn this simple acronym for when they are stranded in hostile terrain with only their jump suits for protection.

C Keep clothing **C**lean.
O Avoid **O**verheating.
L Wear clothing **L**oose and in layers.
D Keep clothing **D**ry.
E **E**xamine clothing for defects in wear.
R Keep clothing **R**epaired.

ROYAL MARINES TIPS

CARING FOR YOUR BOOTS

As they have to 'yomp' everywhere on foot, often over long distances, Britain's Royal Marines have tried and tested rules for the care of boots.

- Stuff wet boots with newspapers and dry them in a warm, airy place, though not in direct heat, which will bake and then crack the leather.
- In winter rub silicone or wax over the laces to stop them freezing when they get wet.
- Your boots should be of a size that allows you to wear two or three pairs of socks underneath.
- Socks which are too tight will restrict the circulation and the layer of warm air that is between them – this can lead to frozen feet.
- Always carry a spare pair of socks.
- Whenever feet get wet change socks as quickly as possible. Boots should also be dried as best they can.

FOOTWEAR

For any outdoor activity it is best to equip yourself with a pair of waterproof boots. Training or tennis shoes should *NOT* be worn; they will not protect you from the cold and wet. The best kind of footwear for general backpacking is walking boots, which have a flexible sole with a deep tread.

It is important to look after your boots, and it is always wise to carry a spare pair of laces around with you. Keep the uppers supple and waterproof with a coating of wax or polish, and always check your boots before you use them for broken seals, worn-out treads, cracked leather, rotten stitching and broken fastening hooks. Remember, if you look after your boots they will look after you, and there is no reason you won't be able to get up to 10 years of use out of them if you do. Many backpackers wear nylon gaiters over their boots to help keep water out when walking through wet grass and the like.

Socks are another important item of footwear, and most backpackers wear two pairs on their feet for comfort and to prevent blisters. Whether you wear a thin pair and a thick pair, two thin pairs or two thick pairs is up to you, but find a combination that suits you before you undertake any serious walking.

TROUSERS

Windproof trousers are recommended for outdoor use, but they should also be light and quick-drying. Synthetic/cotton gaberdine-type weaves are the best. The better makes of trousers are compact, light and extremely quick to dry, even after being soaked. In addition, they have around five pockets with zips, making them excellent for carrying items securely.

WATERPROOF TROUSERS

These fit over your trousers and should have a side zip to allow them to be put on if you are wearing a pair of boots, or they should be wide enough to allow you to put them on while wearing boots. Be careful they are not tight-fitting, if they are your legs will quickly start to sweat. .

JACKETS

Your jacket forms your outer shell, as such it must be windproof and waterproof. A jacket with a covered zip is best, as this will prevent the wind and wet entering, and will be a backup if the zip fails (zip coverings are usually secured by studs of velcro).

The jacket should have a deep hood which can accommodate a hat, comes up to cover the lower part of the face, and has a wire stiffener (hood drawcords with cord locks are better than the tying variety, especially if wearing gloves).

The sleeves should cover the hands and the jacket should have wrist fasteners. It should also be big enough to accommodate several layers of clothing and allow the flow of air in warm weather. The number of pockets is a personal choice, but you should select a jacket that has at least two on the outside with waterproof flaps and one inside that can hold a map. The jacket should be knee-length and also have drawcords at the waist and hem.

Colour is a matter of choice. Some people prefer the more military looking types in olive green or camouflage. However, while it may be more pleasing to the eye to wear a colour that blends into the surroundings, remember that it will be more difficult for rescue patrols to see you in an emergency. Bright colours, on the other hand, stick out and draw attention to yourself - excellent for a survival situation.

A jacket is one of your most important items of clothing, if not *the* most important. Do not compromise when it comes to buying one.

PAYING A LITTLE BIT MORE AND GETTING A GOOD GORE-TEX JACKET WILL GET YOU A HARD-WEARING ITEM THAT MAY SAVE YOUR LIFE.

GLOVES

There are a host of woollen and ski gloves on the market, but mittens are better for heat retention. However, they can be very awkward if you want to use your fingers. Therefore, wear a thin pair of gloves under your mittens.

HEADGEAR

It is estimated that between 40 and 50 per cent of heat loss from the body in some conditions can occur through the head, therefore it is important to wear something on the head (headgear can also provide protection from the heat in hot weather). Any sort of woollen hat or balaclava will help prevent heat loss, though of course they are not waterproof.

WHAT TO CARRY

In the fight for survival a few key items can mean the difference between life and death. All travellers should therefore anticipate any life-threatening situations they may face and carry equipment that will help overcome them.

The simple rule is: do not carry useless weight. No one would consider carrying bricks and other useless weight in his or her backpack. However, if you have a tent, for example, that is totally unsuitable for the terrain you are in, in effect you are carrying around dead weight. Similarly, why carry around bulky tins of food when you can have lightweight packets of nutritious dehydrated food? You only want those things that will serve, not hinder, you.

THE SURVIVAL TIN

The survival tin (Diagram 1) can be one of your most useful pieces of equipment. If you have the items listed below always at hand, your chances of survival, *REGARDLESS OF THE TERRAIN YOU FIND YOURSELF IN*, will be greatly enhanced. These items of equipment are not expensive or difficult to operate, and they can be fitted into an ordinary tobacco tin.

Get used to carrying the tin around with you at all times (it can easily fit into most jacket pockets), and regularly check its contents for deterioration, especially the matches and tablets. Pack the contents with cotton balls or cotton wool - it stops annoying rattling and can be used for making fire.

Your survival tin (A) should include the following items: matches (B), but use only when other improvised fire-making methods fail; candle (C), both a light source and useful for starting a fire (tallow wax can be eaten in an emergency); flint (D), ensure you have a processed flint with a saw striker - this combination can make hundreds of fires, and will carry on working long after your matches have been used up; sewing kit (E), useful for repairing clothes and other materials; water purification tablets (F) are useful when water supplies are suspect and you do not have boiling facilities; compass (G), a small button,

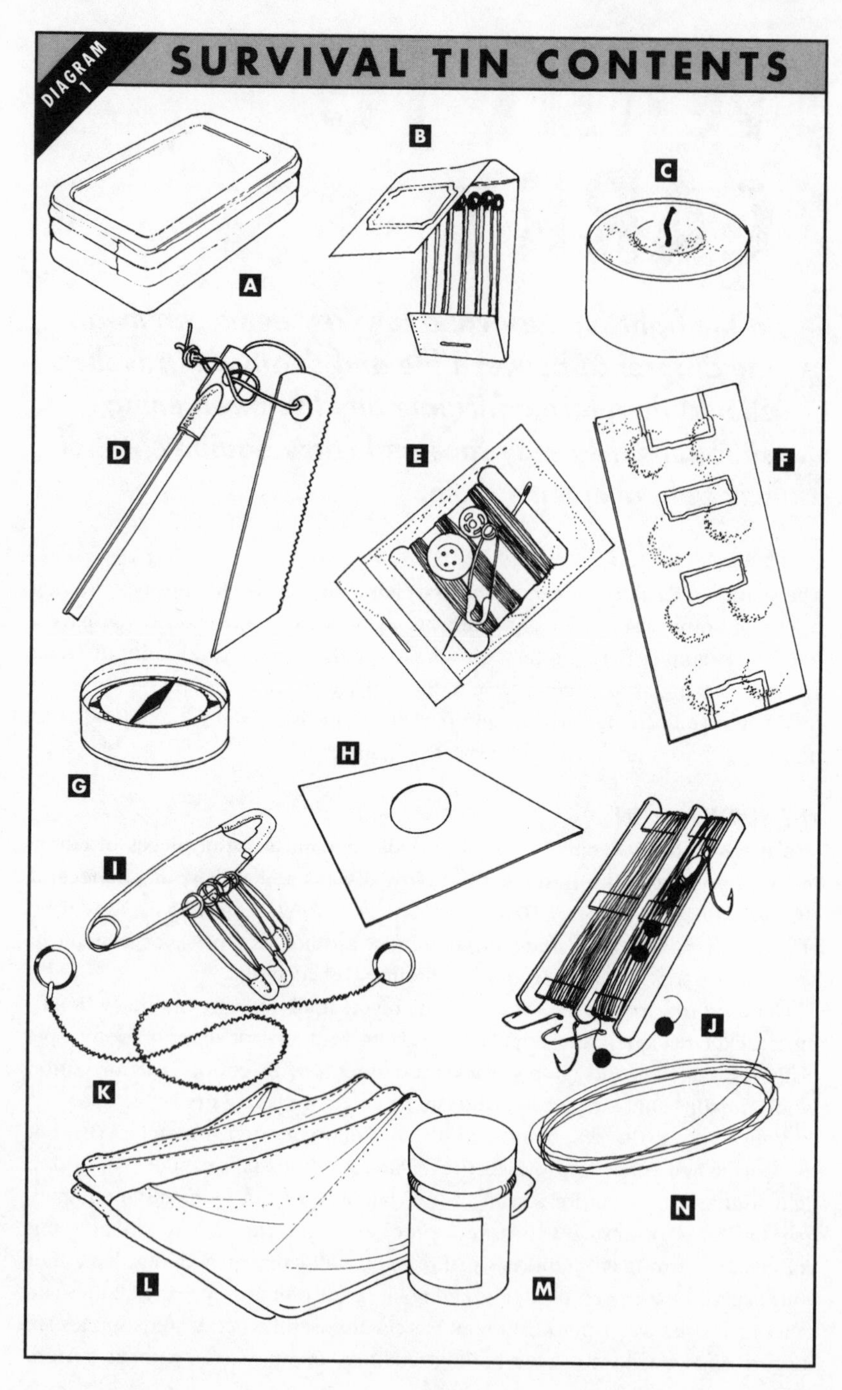
DIAGRAM 1
SURVIVAL TIN CONTENTS
A
B
C
D
E
F
G
H
I
J
K
L
M
N

liquid-filled compass is the best, but check regularly for leaks; mirrors (H) can be used for signalling; safety pins (I), useful for securing items of clothing and for the manufacture of improvised fishing lines; fish hooks and line (J), the fishing kit should also include split lead weights; have as much line as possible - it can be also be used to catch birds (see Food chapter); a wire saw (K) can cut even large trees (cover it in a film of grease to protect against rust); large plastic bag (L), can be used to carry water and also for use in a solar still and vegetation bag; potassium permanganate (M) many uses, for example it can make an antiseptic and treat fungal diseases when added to water; snare wire (N), brass wire is the best, can be used repeatedly for animal traps.

SURVIVAL BAG

It is also useful to make up another, larger survival kit, one that will fit into a small-sized bag and which can be carried in your car or with you on trips. As with the survival tin, get used to always having it with you, and make sure you regularly check its contents for any signs of deterioration.

The items you should carry in the bag are: sewing kit, pliers with wire cutter, dental floss (for sewing), folding knife, ring saw, snow shovel, signal cloth (at least three foot by three foot), fishing hooks, flies, weights and line, three large safety pins, 45m (150ft) of nylon line, gaff hook, multi-vitamins, protein tablets, large chocolate bar, dried eggs, dried milk, file, cutlery set, three space blankets, compass, signal mirror, four candles, micro-lite flashlight, extra battery, extra bulb, fire starter, windproof and waterproof matches, butane lighter, flint, insect repellent, 12 snares, spool of snare wire, tin opener, plastic cup, water purification tablets, sling shot and ammunition, knife sharpener, whistle, soap, two orange smoke signals, 67m (225ft) of nylon twine, 67m (225ft) of nylon cord, one pair of work gloves, a mess tin and a mouse trap.

> **WARNING**
>
> ***Do not cheat on quality when it comes to items for your survival kits, you could be gambling with your life. In addition, do not just make up kits and forget about them, check them regularly for any signs of deterioration.***

TENTS

A portable shelter is an essential item of any outdoors kit. As with the clothing, there is a vast range to choose from, ranging from ultra-lightweight mountain and arctic models to cheap and cheerful summer weather types. To select a tent that is tailored to your specific needs, collect brochures and magazines, go to tent displays or to camping shops that have them displayed on the premises. Most modern tents are not heavy, so unless weight is your overriding priority, go for a tent that has plenty of internal space.

Bivi-bags are portable shelters that have gained popularity in recent years, In reality a waterproof overall for a sleeping bag, some have hoops which convert them into a low-profile, one-man tunnel tents. You can't really cook inside a bivi-bag, but it is windproof, waterproof and very lightweight - as little as 0.54kg (19oz). In addition, because they are 'breathable' condensation will not collect inside them, thus your sleeping bag will remain dry.

Frame tents Geodesic (dome-shaped) tents offer plenty of internal space. In addition, many have enough space between the fly sheet and the inner tent to allow you to store equipment or to cook. If they have two entrances then you can do both - which keeps the inner tent relatively free of clutter. Another factor that may attract you to a tent is the provision of mosquito nets that can be zipped over the entrances, this will be God-send in summer, especially if you are camped near a water source.

SPARE CLOTHING

There are two aspects to this. First, the survival clothing you carry with you in your car or on an aircraft journey when you are wearing light, everyday clothes. Second, the spare clothing you carry with you in your pack when you are back-packing.

In the first instance, you should carry those items of clothing that were listed in the What to Wear chapter (see above). In the second instance, the articles of clothing should be spare socks, underwear, shirts and 'long johns', ie those items of clothing that come into contact with the skin, absorb perspiration and thus get dirty and decay through wear and tear. Outer clothing, including boots, should last for years if they are good quality and properly looked after, and thus it is rather a waste of space and weight to carry spares (though do carry items such as wax for your boots and waterproofing treatment for your tent and jacket). Don't forget to carry spare boot laces.

COOKING EQUIPMENT

There are lots of commercial cooking stoves to choose from, but there are two very important rules you should bear in mind when making your choice:

- Keep it as lightweight as possible.
- Do not buy a stove that has a lot of extra bits - they can be easily snapped off and lost when being used in the outdoors..
- As with everything, you must choose the stove that fits your needs, though aim for one that weighs between 500g and 700g. The choice of fuels for stoves is also wide: butane/propane, methylated spirit, paraffin and petrol. However, if you intend to cook inside a confined space note the following:
- Paraffin cookers should only be re-fuelled when they are cold. When they are burning, ventilate the tent to prevent the build-up of toxic gases.

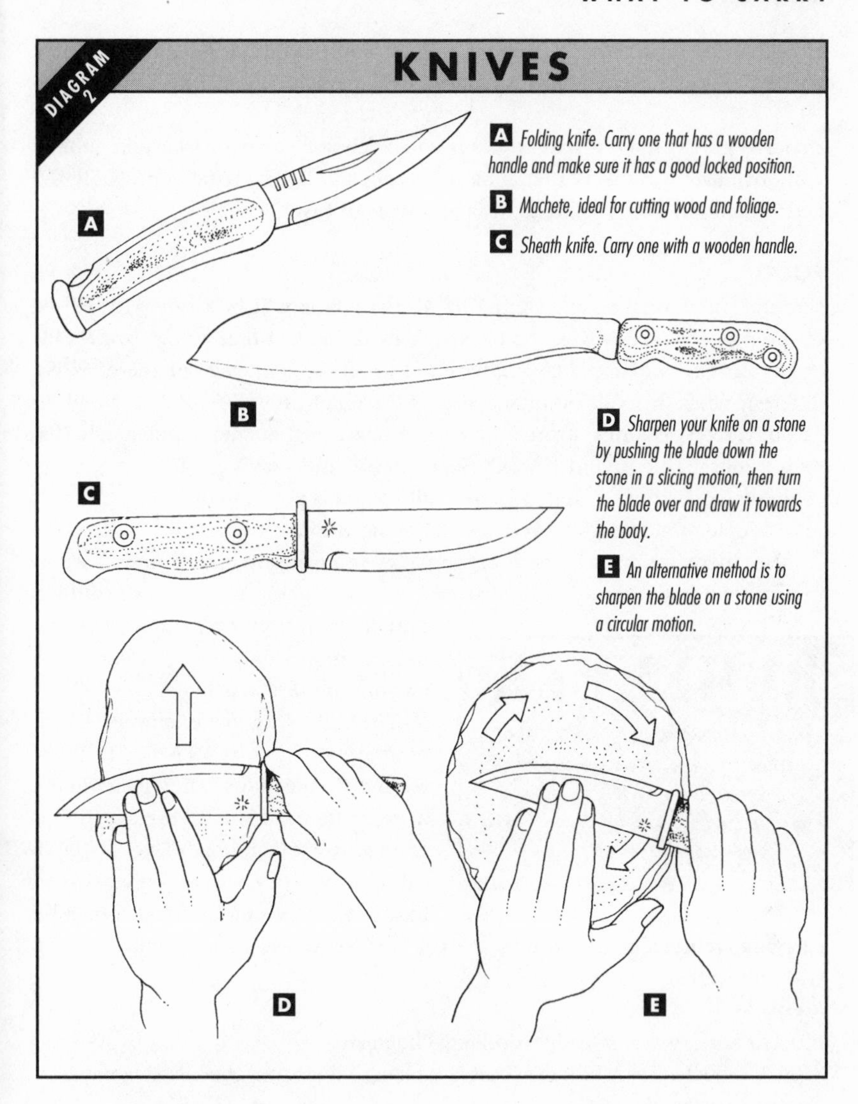

- Gas can freeze in low temperatures.
- Leaded petrol in cookers is a danger to health when burnt inside a tent; use de-leaded petrol, called Naptha, or white petrol.
- Hexamine blocks must never be burnt inside a tent.

Cookware There is a large choice available, ranging from the ubiquitous aluminium mess tins to stainless steel cook sets. The latter usually come in sets of four or five items that fit inside one another to form a very compact set –

excellent for space saving. However, before you rush out and buy a set, ask yourself if you really need this amount of cooking tins.

Crockery As with cookware, there is a wide choice currently available, though you obviously want a set that is hard-wearing and light. Plastic is probably the best; as it fulfils both requirements and it will not rust.

FOOD

General nutritional requirements for the survivor will be dealt with in the chapter on food. However, the backpacker, having had time to prepare for his or her journey, will hopefully not have to live off the land, and will therefore eat different types of food. He or she should carry dehydrated food that is high in carbohydrates, calories, protein and has a degree of vitamin supplementation. As a general rule, avoid tins, they are bulky and awkward to carry. Many camping foods come in sealed sachets, all you do is pour in hot water, stir, wait for a couple of minutes, and you have a hot, nutritious meal.

Also, there are many survival ration packs you can purchase, compact packs that contain high-calorie meals that can keep you going for 24 hours (though they keep you going for longer if you use them sparingly and you cut down on your physical activity.

> **REMEMBER**
>
> ***Whatever food you carry, always have emergency rations stowed untouched as well. Even if it is only raisins, biscuits, chocolate or compact energy food – it will keep you going for 24 hours.***

The following are examples of 24-hour arctic ration packs issued to Royal Marine Commandos. They guarantee a minimum of 4500 calories a day, enough to sustain a high level of physical activity. They will also give you an idea of the types of food you should be carrying (remember that the meat, fruit and vegetable meals require mixing with hot water):

Menu A:

BREAKFAST – warm porridge, drinking chocolate.

SNACK – beef spread, biscuits (fruit and plain), chocolate, chocolate caramels, nuts and raisins, dextrose sweets.

MAIN MEAL – chicken soup, beef granules, mashed potato powder, peas, apple flakes.

Menu B:

BREAKFAST – warm porridge, drinking chocolate.

SNACK – chicken spread, biscuits (fruit and plain), chocolate, chocolate caramels, nuts and raisins, dextrose sweets.

MAIN MEAL – vegetable soup, curried beef granules, rice, peas, apple and apricot flakes.

Menu C:
BREAKFAST - porridge, drinking chocolate.
SNACK - chicken and bacon spread, biscuits (fruit and plain), chocolate, chocolate caramels, nuts and raisins, dextrose sweets.
MAIN MEAL - oxtail soup, mutton granules, mashed potato powder, peas, apple flakes.
Menu D:
BREAKFAST - warm porridge, drinking chocolate.
SNACK - ham spread, biscuits (fruit and plain), chocolate, chocolate caramels, nuts, raisins, dextrose sweets.
MAIN MEAL - vegetable soup, chicken supreme granules, rice, peas, apple and apricot flakes.

KNIVES

A knife is extremely important in a survival situation. It can be used for many things, such as skinning animals, preparing fruits and vegetables and cutting trees. Therefore, always keep your knife clean and sharp and make sure it is securely fastened when you are travelling. There are many knives to choose from (Diagram 2), but it is best to have one that has a single blade and a wooden handle.

NEVER THROW YOUR KNIFE INTO TREES OR INTO THE GROUND, YOU COULD DAMAGE OR LOSE IT.

BACKPACKS

They are many backpacks available to the adventurer, ranging from small 20-litre capacity packs to the large 100-litre capacity bergens. However, remember to get a pack that is suited to your needs. If you get a 100 litre pack when really you only need a 50 litre one, you will probably end up filling the large pack to the brim and thus carry around a lot of unneeded weight. There is a fine line between necessity and excess; you must learn to recognise it.

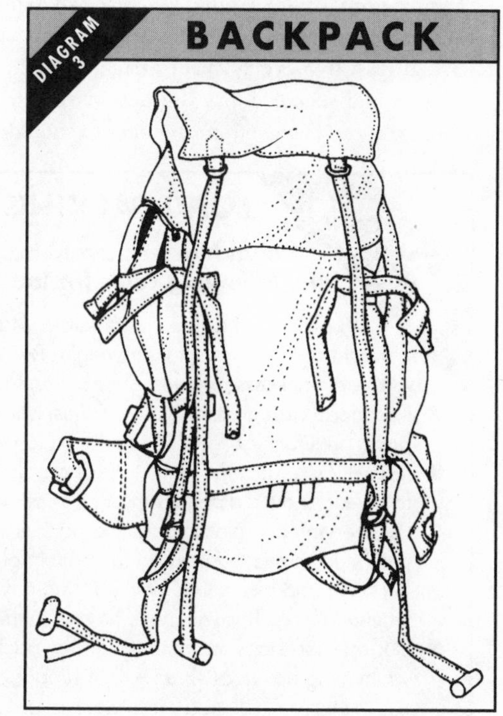

Backpack design has advanced in recent years, and the H-frame pack has, to a certain extent, been outdated by modern packs with anatomically shaped shoulder straps, hip belts and lumbar pads with aluminium frames (Diagram 3). That said, if you are going to carry heavy loads for long periods, then it is better to have an H-frame pack. If you decide to stick to a contoured pack with an internal frame. It is important that your pack feels comfortable to wear - people come in all shapes and sizes and so do backpacks. Here are some of the things you should look for when buying a backpack:

- ☐ Side pockets, useful for carrying items that you need ready access to.
- ☐ Side compression packs, useful for evenly distributing the the load inside the pack and for carrying additional equipment.
- ☐ Base compartment, allows the load in the pack to be divided for weight distribution purposes and ease of access.
- ☐ Extendible lids, can vary backpack capacity.
- ☐ Double stitching, binding and bar tacks increase strength and protection.

Horseshoe pack (Diagram 4) At the other end of the backpack scale is the horseshoe pack, which can be used to carry items comfortably over long distances. The procedure is as follows: lay a square-shaped material 5ft x 5ft (1.5m x 1.5m) on the ground (A), place all items on one side of the material and then roll the material with the items to the opposite edge. Tie each end and have at least two evenly spaced ties around the roll (B). Then bring both ends together and secure. What you now have is a compact and comfortable pack, which you can change from shoulder to shoulder if required (C).

ROYAL MARINES TIPS

LOAD PACKING AND CARRYING

Royal Marines avoid back injuries by observing the following rules for load carrying.

- ■ Keep load as light as possible. Maximum load per person should be a quarter of their weight. Resist the temptation to fill your pack with unnecessary bulk.
- ■ Keep load as high as possible. Adjust packs to keep load close to the back, but DO NOT restrict circulation to the arms.
- ■ Inside packs, arrange items to give a balanced loading. Corners of tins, footwear and hard objects must be kept away from the back.
- ■ Put everything in plastic bags (no pack is 100 per cent waterproof), and put the least-needed items at the bottom of the pack.
- ■ Put stove and fuel in side pockets, and anything else that will be needed when walking. This will make having to take off the pack unnecessary.
- ■ During short stops, do not take off the pack; rather, use it as a back rest when lying down, or sit up with it supported on a rock or log.

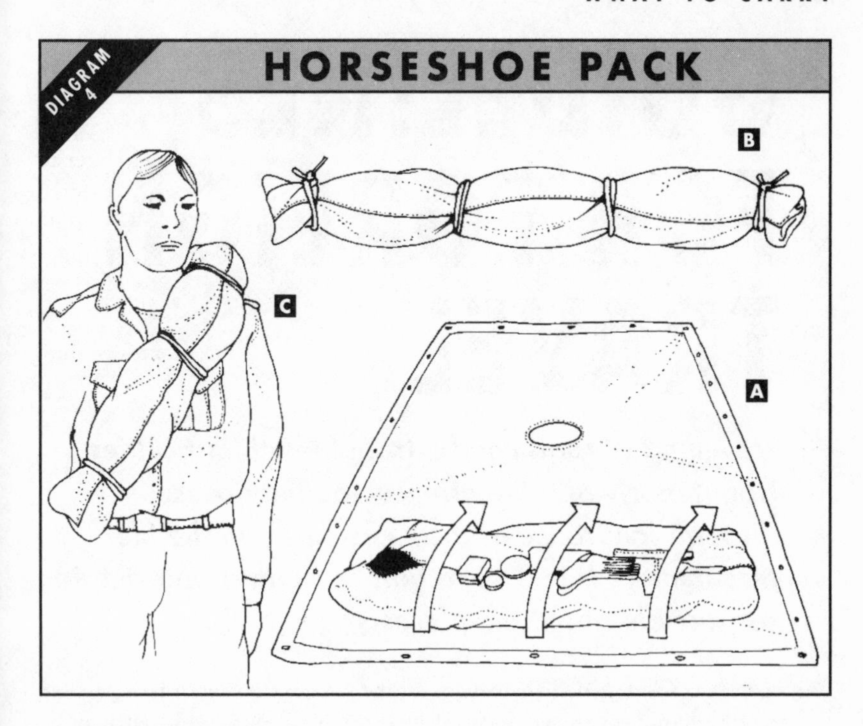

SLEEPING BAGS

Good sleeping bags are filled with down, the best insulating material. In wet conditions you will need a waterproof cover for a down-filled bag. If you know you will be sleeping in wet conditions then you should get a bag filled with Holofil. You could opt for an all-weather sleeping bag, which are light but expensive. They consist of a sleeping bag, a fleece liner and a bivi-bag.

BRITISH SAS TIPS

BASIC SAS MEDICAL KIT

SAS medical kits include treatments for breathing and circulation difficulties, bleeding, breaks and fractures, burns, infections and pain.

- Paediatric mucus extractor.
- Blood volume expander.
- Infusion fluid.
- Dressings.
- Artery forceps.
- Suture kit.
- Fracture straps.
- Burns dressings.
- Antibiotic tablets.
- Antibiotics in injectible form.
- Pain killers.
- Flamazine cream.
- Creams for fungal infections.

GUIDELINES FOR FOREIGN TRAVEL

Travelling abroad can be fraught with difficulties, from the trivial – the offence you may cause through ignorance of local customs – to the life-threatening. It is wise to gather as much information as possible about the places you will visit.

This chapter contains general guidelines for the foreign traveller. It is impossible to provide a comprehensive account of every local difficulty that may be encountered abroad, but you can make sure that you have learned as much as possible and take every reasonable precaution. In particular, you are advised to stay well clear of the trouble spots listed below.

As with survival skills in general, the traveller is strongly advised to do some prior research on the countries and areas he or she is planning to travel to. There are some excellent publications on the market, such as the Lonely Planet guides, Cadogan Guides and the books published by Vacation Work Publications. These will tell you when to go, what to visit, what inoculations to have and where to avoid. With all this information readily and relatively cheaply available, there is really no excuse for not finding out about your destination before you arrive.

BEWARE

Some countries will refuse you entry if you have certain stamps in your passport. For example, many African countries will refuse you entry if you have been to South Africa, and the Arab countries of North Africa (except Egypt) will refuse entry to Israeli nationals and anyone with an Israeli stamp in their passport. You should plan your movements through Africa and the Middle East before you set off.

ESSENTIAL DOCUMENTS

You must have a passport for foreign travel. Try to ensure that it is valid for a reasonable length of time (many countries insist on your passport being valid

WARNING

AIDS is widespread throughout Africa, the Middle East and the Far East. Do not share needles if injecting drugs, and beware of immunisations at Third World border posts (they are often carried out en masse with one needle, and there is no attempt to sterilise it). Do not indulge in unprotected sex or anal sex. Use a condom (though remember that they are not 100 per cent effective), and stay away from prostitutes, especially in the Third World.

for at least six months from the date of entry into their land, regardless of your actual length of stay). In addition, make sure it has plenty of blank spaces: many officials insist on rubber-stamping foreign passports. If you are crossing a lot of borders, your pages can be quickly used up.

Make a note of your passport number and its issue date, and make photocopies of the relevant data pages. This will be a godsend if you happen to lose your passport.

With regard to visas, check beforehand to see if you need a visa to gain entry into a specific country. Do not turn up at a border without a visa unless you are certain you can get one there: you may be sent back to your nearest consulate, which in a large country can be hundreds of kilometres away. If you know you will need a lot of visas, make sure you have a large amount of passport-sized photographs for the application forms.

An International Vaccination Card is also considered essential for foreign travel. You may consider one unnecessary, but remember that if you don't have one some countries will insist you have the necessary vaccinations at the border. As the use of a sterile needle cannot be guaranteed, you run the risk of contracting AIDS. Something to think about!

RESPECT OTHER CULTURES

Wherever you go in the world, bear one thing in mind: it is not your country. You should respect the indigenous population's customs and religions, even if you do not hold with them, and always treat religious artefacts, places and ceremonies with respect. In addition, try to find out about local customs and religious taboos before you go to an area, so as

DRUGS

DON'T EVEN THINK ABOUT IT!

Many countries, especially in the Far East, have stiff penalties for possession of illegal drugs and for drug trafficking. They range from long stretches in appalling prisons to the death penalty. In addition, many Third World 'drug dealers' often turn out to be plainclothes policemen trying to catch drug dealers and users. In some countries you can be sent to jail for being associated with someone who has been found with drugs on them, so keep an eye on your companions. Do not be under the illusion that foreign states will treat Westerners more leniently for drug-related crimes – they won't. In fact, the opposite might be the case, with the authorities seeking to make an example of you. If you get caught you could be hanged or sent to prison for life. You have been warned.

not to offend. That goes for codes of dress and social etiquette, too. In this way your trip is more likely to be trouble-free and enjoyable.

AFRICA

Photography Many African countries have restrictions on what can and cannot be photographed. As a general rule, do not take photos of anything connected with the military, including bridges, railway stations, post offices, radio and television stations, prisons and port facilities. You will need a photography permit for the following countries: Angola, Benin (the authorities are suspicious of cameras, better to be safe than sorry), Burkino Faso, Cameroon, Chad, Equatorial Guinea and Somalia.

Respect people's privacy. Many locals do not like having their photos being taken, especially some tribal people. Others will insist on a small fee before allowing themselves to be photographed. Before you start taking offence at this, remember you are in their country and that they are very poor.

Hitchhiking is a recognised form of public transport in Africa (in some instances you will have to pay a fee), though remember that sticking out your thumb is an obscene gesture in many African countries.

Customs and religions The adventurer should be aware of the following customs and beliefs when travelling through Africa:
ALGERIA: the south of the country is very traditional. Women, for example, are regarded as chattels, and any woman over the age of 12 who isn't veiled or locked up is regarded as a whore, or at the very least sexually available (an opinion that is common throughout the Arab world).
ANGOLA: the province of Lunda Norte, in the northeast of the country, is out of bounds to foreigners.
BOTSWANA: do not tangle with the Botswana Defence Force (the army); you will be arrested if you go near any military establishments. It is considered very impolite not to greet people that you meet.
BURKINO FASO: expect to encounter many roadblocks and line-ups at checkpoints at gunpoint. DO NOT go to help locals who are being beaten up by police officers for being untidy and 'disrespectful'; if you do you will get the same treatment. Be polite and docile.
CONGO: avoid talking politics with strangers. The government employs plainclothes police in the towns and cities to curb 'subversion'.
GABON: ensure you have all your papers at all times: the police like to hassle foreigners, especially whites. You will probably lose your camera if you are seen taking a picture of a military installation.
KENYA: because of tourism, the capital Nairobi is infested with thieves and con artists. Keep your wits about you if you visit there.

LIBYA: be careful with cameras - most Libyans are suspicious of foreigners with cameras, largely as a result of government-controlled media scare stories.
MALAWI: it is illegal for women to wear trousers or skirts that don't cover their knees. Men with long hair or those wearing shorts without long socks, or wearing flared trousers, are barred from entry.
NAMIBIA: avoid the presidential summer palace in Swakopmund when in residence - the Presidential Guard is very trigger-happy! Sperrgebeit is a prohibited diamond area. The heavily armed guards who patrol it shoot first and ask questions later.
NIGERIA: the capital Lagos is a high-crime area. Smoking anything in public places is illegal, and there is a mandatory death sentence for drug trafficking.
SIERRA LEONE: the traditional handshake is with the right arm with the left hand grasping the right wrist or forearm.
SOUTH AFRICA: you are likely to encounter much racism if you are not white, and sexism is endemic among all races.
SUDAN: do not bring in radios, expensive cameras, videos, batteries and foreign newspapers and magazines into the country. They will be stolen or confiscated. Alcohol is illegal under Islamic law and if you are caught in possession of it you may be fined and flogged.
ZAIRE: there is intense paranoia about spies throughout the country, so be wary of giving the impression in any way that you are one. Stay away from the diamond mining towns of Mbujimayi and Tshikapa: any travellers found near them are assumed to be diamond smugglers.
ZAMBIA: to take photographs of anything military or official is to invite the police or army to give you a hard time - they will think you are a South African spy! It is best to keep a low profile. In addition, violent robbery is spiralling throughout the country. Do not go out at night or display signs of wealth.
ZIMBABWE: as a result of the dire state of the economy and the poverty of the people, theft and street crime are increasing alarmingly throughout the country, especially in Harare and Bulawayo.

THE MIDDLE EAST

Photography The same rules apply in the Middle East as in Africa: do not photograph anything remotely military in nature. In addition, remember that few Middle Eastern officials will be pleased if you take pictures that show poverty in their country and therefore give a bad impression to outsiders.

Women travellers Because of the conservatism of some Middle Eastern states, women travellers can face difficulties when travelling alone. To avoid getting harassed it is wise to stay in good hotels. Do not flirt with strange men or make eye contact with them. Dress conservatively (cover your arms and legs, sometimes also the head, and do not wear anything that is tight-fitting) and do not ride in the front seat of taxis. Check up on the laws governing women before you go to an individual country.

Ramadan This is the month during which Moslems fast from dawn until dusk. Everyone, regardless of their religion, is required to observe the fast in public, and penalties for breaking it can be severe, especially in Saudi Arabia. Be warned. In addition, if you are travelling in an Islamic country take note of the festivals and holy days. It helps when planning trips and shopping.

Customs and religions In general, always stand when someone enters the room. When you enter a room always shake hands with everyone, touching your heart with the palm of your right hand after each shake. Do not offer to shake an Arab woman's hand unless she extends her hand first. When two men meet it is polite for each to enquire after the other's families, but *not* each other's wives. When you are seated do not do so in a way that points your soles at someone, it is very impolite. It is also impolite to refuse an offer of coffee or tea in any social or business situation.

One more thing: if you are a Westerner do not adopt traditional Arab dress, the locals will assume you are mocking them. When visiting mosques (if you are allowed), dress conservatively, be polite and take off your shoes.

BAHRAIN: as in many places in the Middle East, do not take any photographs of Arab women.

KUWAIT: walking in the desert is extremely risky until all the mines laid during the Gulf War are cleared, and beware of mines washing up on the beaches. Do not venture north near the Iraqi border: it isn't marked and you could wander into Iraq by mistake. If you do, you will be arrested and probably put on trial in Baghdad for spying.

OMAN: a very security-conscious society. Observe extreme caution when taking photographs.

SAUDI ARABIA: strictest of the Arab countries. Women are not allowed to drive and must be accompanied by their husbands or a male relative when travelling by bus or train. Alcohol is illegal. Do not attempt to smuggle it into the country, you could get jailed or even flogged for doing so. Avoid the Religious Police, the *Matawwa*. They take their work extremely seriously.

TURKEY: there are restrictions on areas bordering the former Soviet Union and Kurdish regions where there is unrest. Do not attempt to go to these areas without permission.

ASIA

Women travellers To avoid unwanted attention from men, you might consider wearing a walkman and dark glasses, although some countries regard this as a sign of a loose woman! However, it does get round the problem of making eye contact with strange men. The general rule is: if you are a woman do not travel alone. Loose-fitting clothes are essential (they are cooler anyway), and you may like to wear a wedding ring to deter male pests.

Customs and religions Cutlery is rarely used when people eat. Thus washing your hands before and after meals is a necessary ritual. In addition, only the right hand is used for eating and handing things to others. The left hand is considered unclean (many Asians still use it instead of toilet paper). Get into the habit of using your right hand.

If you visit a temple or shrine, remove your shoes or sandals before entering. This also applies to entering a private house. You may also have to wash your feet before entering some temples. Cover your arms, legs and head when entering Sikh temples or Moslem mosques. Women must not come into physical contact with Buddhist monks: if you want to give a monk something put it on the ground first.

Couples should generally act more soberly than they do in the West, thus no holding hands or kissing in public. In addition, touching someone's head is considered offensive in some Asian cultures.

One more thing. Asians will generally go out of their way to be polite and not to give offence. Thus some of your questions and enquiries may elicit inaccurate answers. The person who gives you the information has not deliberately lied, merely attempted to be polite or not give you bad news. The best way to get round the problem is not to ask questions that require a simple 'yes' or 'no' answer, but require a more detailed response.

IRAN: contrary to popular opinion, Westerners are not hanged from the nearest lamppost. That said, anti-Western propaganda, particularly against the United States, has had an influence, so the advice is: be cautious. Keep your arms and legs covered, and if you are a woman wear a headscarf.

AFGHANISTAN: avoid this war-ravaged country. You cannot get a tourist or transit visa, and even if you do get in there are curfews in the towns and cities, the hotels and restaurants are shut and transport services are severely limited. In addition, the countryside is littered with mines left over from the war between the Mujahedeen and the Russians.

PAKISTAN: be careful about dress, especially women. It is advisable to wear the traditional *shalwar komeez* (baggy trousers and long shirt/dress). Exercise extreme caution when taking photographs of tribal people, especially the wives. You could get yourself into a lot of trouble with their husbands.

INDIA: beware of thieves and robbery in general. Drugging of travellers for the purposes of theft is on the increase.

BANGLADESH: after Saudi Arabia, the country that most strictly adheres to Ramadan. Be very careful not to offend by the way you dress, especially be careful not to show the soles of your feet.

THE FAR EAST

Photography Apply the same rules as in Africa (see above)

Appearance Generally speaking, officials in the Far East have an intense dislike of hippies and anyone else they consider to be 'freakish'. Clothes that are particularly disliked include thongs, shorts, jeans and T-shirts. If you look neat and affluent your trip will be much less troubled when dealing with officialdom (though don't look too affluent - you will draw the attention of con men, pickpockets and thieves!).

Customs and religions The adventurer should be aware of the following rules and customs when travelling through the Far East:
BURMA: the 'Golden Triangle' and near the Chinese border are no-go areas.
CAMBODIA: do not travel outside those areas that are under government control. Do not touch any rockets, artillery shells, mortar mines or other military material you come across - it may have just been laid.
CHINA: theft is endemic in this massive country: be on your guard.
LAOS: the northwest of the country is an area of heroin production and smuggling and should be avoided.
THAILAND: do not criticise the Thai royal family: you could end up in jail. Avoid the areas along the Burmese and Cambodian borders.
VIETNAM: do not stray from well-travelled roads and paths, you may step on a mine or other military ordnance left over from the war with the US. Public violence is common. Small foreigners, especially women, often get picked on.

THE AMERICAS

Photographs It is extremely unwise to take photographs of anything to do with the military. Apply the same rules concerning photography in Central and South America as in Africa and the Middle East (see above).

Appearance In Central and Latin America, where there have been numerous military coups and repressive regimes, many people loathe and fear the military. It is therefore not advisable to associate yourself with them by equipping yourself with clothing from your local army surplus store.

In general, dress down so as not to draw attention to yourself. Do not wear expensive jewellery or carry valuable cameras. Remember that petty crime, largely because of economic hardship, is rife.

Identification Always carry identification papers. You will invariably be subjected to a number of spot checks by the military and police during your trip. Be patient, they are often carried out in an effort to curb terrorism and drug-related violence, so they are to your advantage.

THE WORLD'S TROUBLE SPOTS

There is only space here to make the briefest reference to the conflicts that are going on around the world. It is best to avoid the affected areas if possible, as you risk imprisonment under very harsh conditions, serious injury or death. Forget any myth of it being an adventure to venture into war zones. Wars mean death, and could include your own.

Africa

ALGERIA: foreigners targeted for attacks by Islamic fundamentalists.
ANGOLA: wracked by civil war until 1991, when an uneasy truce was signed. Continued fighting throughout the whole country is a certainty. The standard of living is low and public safety is very poor. *AVOID*
BURUNDI: inter-tribal friction in the north of the country.
CHAD: one of the poorest countries on earth. Conflict between the Moslem north and the Christian and animist south. Main trouble spots are the borders with Cameroon, Libya and Sudan. *AVOID*
LIBERIA: inter-tribal civil war. There is complete anarchy throughout the country and no end in sight. *AVOID*
MOROCCO: the conflict between the government and guerrillas in the Western Sahara still simmers.
MOZAMBIQUE: civil strife between the government and guerrillas has supposedly ended, but there is still tension throughout the country.
SOMALIA: the country has collapsed since the start of the civil war in 1988. Despite UN intervention, the only rule is that of the gun. *AVOID*
SOUTH AFRICA: crime is soaring and there is a high murder rate, especially in the townships. Despite the ending of apartheid, the country appears to be on the verge of a major disaster. If you are white stay out of the townships.
SUDAN: civil strife between the Moslem north (the government) and the largely Christian south (the rebels). Travel is very dangerous in the south.
UGANDA: despite the cessation of hostilities between the government and rebels, the army is committing abuses in the north and east of Uganda, and there are gangs of armed bandits roaming the country. *AVOID*

The Middle East

LEBANON: the statistics speak for themselves: in a country with a population of under three million, there are over 100,000 armed troops active within its borders, representing various terrorist groups and foreign armies. Westerners could face death, kidnapping or injury. *AVOID*
IRAQ: country under the severe repression of the dictator Saddam Hussein. Anti-Western feelings are very high as a result of the 1991 Gulf War. *AVOID*
ISRAEL: avoid the Gaza strip and the West Bank, areas of high violence between Palestinians and Israeli settlers and troops.

Asia

AFGHANISTAN: despite Soviet withdrawal, civil strife continues. *AVOID*

BANGLADESH: guerrilla conflict against the government in the Burmese hills in the east of the country.

INDIA: there are guerrilla movements active in Manipur, Nagaland (between Assam and Burma), West Bengal, the Punjab and in Kashmir.

The Far East

BURMA: there are an estimated 28 insurgent groups active within Burma. Areas to avoid are the Thai, Indian and Chinese borders. The country is under the control of the military, but could explode into civil insurrection at any time.

CAMBODIA: the civil war has ended but the country is destitute. There are still guerrilla bands roaming the country and a multitude of unexploded mines and ordnance litter the countryside.

The Americas

BOLIVIA: a major cocaine production centre. Production is concentrated in the Chapare region: do not wander off the Cochabamba-Santa Cruz highway, the area around it is in the hands of drug barons.

BRAZIL: crime and violence, especially in the towns and cities, are rife. Do not wear or carry anything that is a sign of wealth.

COLOMBIA: theft is widespread, especially in Bogota and Cartagena. Kidnapping is an everyday event and foreigners are often a target. Dress down. Anyone who trespasses onto the drug trade, be it wandering into a field of opium poppies or a drug factory, will almost certainly be shot. Avoid in particular the highway between the Venezuelan border and Riohacha.

EL SALVADOR: continuing low-level violence between government forces and guerrillas throughout the country.

NICARAGUA: Contra and Sandanista violence could erupt at any time.

PERU: around three-fifths of the country are controlled by Marxist guerrillas. Avoid in particular the upper Huallag valley and the area from just south of Cajamarca in the north down to the Abancay area in the south. Keep away from all emergency zones and do not travel at night.

Europe

ARMENIA: ethnic violence between Azerbaijanis and Armenians.

AZERBAIJAN: ethnic violence between Armenians and Azerbaijanis.

GEORGIA: ethnic violence between Georgians and Abkhazians.

YUGOSLAVIA: the country has ceased to exist following the ethnic and religious violence between Serbs, Croats, Bosnians, Macedonians, Christians and Moslems. Civil strife in Bosnia set to continue indefinitely, and could spread to other areas of the region. *AVOID*

HOW TO GET OUT OF TRICKY SITUATIONS

A survivor must know how to get out of potentially dangerous situations. In many ways this can be done by anticipating such situations and taking action to ensure you don't get into them in the first place. Thus, you can avoid a lot of trouble if you don't go to a country that has a high level of violence or is in a state of civil war; you can avoid being mugged by not going out alone at night in a bad neighbourhood; and you can stop yourself being robbed in broad daylight in a poor Third World country by not openly displaying signs of wealth. A little forethought can save you a lot of trouble.

Getting out of trouble abroad When dealing with officials, soldiers and policemen abroad try to stay calm and polite at all times. Getting angry and ranting and raving will do you no good, and will probably make the situation worse. Many Third World soldiers and policemen are not particularly well trained or educated, and their tolerance level may be very low.

If you are arrested, ask to be put in touch with your embassy or consulate, stay calm, and ask to see someone of a high rank who can speak English.

If you are unfortunate enough to be in a country where large-scale violence flares up, observe the following rules:

☐ Get off the streets.
☐ Stay away from potentially sensitive places, such as radio and television stations and airports.
☐ Sit tight in your hotel.
☐ Join, and stay with, a group if you can.
☐ Never break a curfew.

BRITISH SAS TIPS

GUIDELINES FOR AVOIDING PERSONAL ATTACKS

SAS soldiers are trained to a high level in unarmed combat, but when acting undercover they are also taught how to avoid potentially dangerous situations.

■ Do not use short cuts across wasteland, through underground walkways, along canal paths or through car parks at night.
■ Do not display expensive items of clothing or jewellery.
■ Avoid eye-to-eye contact with strangers as you pass, though still attempt to assess their intentions.
■ Do not appear nervous, keep your head up and appear confident (even if you are not).
■ Do not let anyone who approaches you get within arm's length.
■ Cross the road if you see a group of people approaching you.

Getting out of tricky situations at home The main weapon you possess that can save you from being attacked is your common sense. For example, never go out for the evening without planning how you will get back home, do not use short cuts across wasteland or down badly lit passageways at night, and do not display expensive jewellery, bags or clothes. It's all common sense.

Physical attacks are most common on individuals, and occur most frequently during the summer, between mid-evening and early morning, away from other people, from behind, when pubs and nightclubs have just closed, and when people have just been paid (and have their money with them).

Travelling on public transport Using public transport, especially at night, can be risky. However, as with other areas of potential risk, try to use your common sense when travelling on buses, trains and subways. For example, on the subway always choose a compartment that has lots of people in it, and if you feel threatened in any way get off and wait for another train or take the bus. In addition, when waiting for a train, stand on the busiest part of the platform – do not make yourself an easy target. If you do not feel safe using any public transport at night, take a taxi. Do not put yourself in danger unnecessarily for the sake of saving some money.

NEW YORK PD TIPS

GUIDELINES FOR TRAVELLING ON THE NEW YORK SUBWAY

New York is one of the most violent cities on earth, and many attacks occur on public transport. Follow New York's police guidelines for avoiding trouble.

- Do not make eye contact with anyone who appears to be 'looking for trouble'.
- Keep alert. Even if you are reading or wearing a walkman, be aware of what is going on around you.
- If you are standing, do so where you cannot be approached from behind.
- Use the interconnecting carriage doors if necessary to avoid trouble.
- If your carriage empties, move to a crowded one.

If you are subjected to an attack, you will probably have no alternative but to defend yourself as best you can (unfortunately you cannot rely on your fellow passengers to aid you), but if your bag is being snatched do not automatically put up resistance. You may make the situation worse by forcing your attacker to use more violence against you. A mugger may have a knife or a gun and you may be forcing him to use it. *ASK YOURSELF THIS QUESTION: ARE YOUR POSSESSIONS WORTH MORE THAN YOUR LIFE?*

GOLDEN RULES

Regardless of the terrain you find yourself in, the acquisition of water food and shelter will be your main priorities in a survival situation. In addition, you must have a thorough knowledge of wilderness first aid, ropes and knots, and signalling and navigation.

WATER

Of all the elements of survival, water is the most important. Without water there is no life, therefore all survivors must know where to find and how to make water if they are to live in the wild for any length of time.

In a survival situation a person may be able to live without food for weeks in certain conditions. Without water, however, he or she will die within days. Therefore, finding water is the number one priority for all survivors.

REQUIREMENTS

In a temperate climate a person needs to drink at least 2.5 litres (0.5 gallons) of water a day. Even if you are doing very little physical activity, you need water to replace fluids lost through the following:
URINE: approximately 1.5 litres (0.33 gallons) of water lost per day.
SWEAT: approximately 0.1 litres (0.022 gallons) of water lost per day.
FAECES: approximately 0.2 litres (0.044 gallons) of water lost per day.
DIFFUSION THROUGH THE SKIN: the diffusion of water molecules through the cells of the skin. Around 0.4 litres (0.088 gallons) of water lost per day.
EVAPORATION THROUGH THE LUNGS: when air enters the lungs it is usually fairly dry. However, in the lungs it comes into contact with the fluids covering the respiratory surfaces, which saturate it. It is saturated with water when it leaves the lungs.

The importance of drinking water on a daily basis can therefore be seen. In addition, water requirements are increased if you suffer any of the following:
HEAT EXPOSURE: when exposed to high temperatures, an individual can lose up to four litres (0.88 gallons) of water per hour through sweat.
EXERCISE: increased water loss through the lungs is a result of a higher respiration rate and increased sweating, both the consequences of physical exercise.
COLD EXPOSURE: the amount of water vapour in the air decreases and the temperature falls in the cold. Breathing in cold air results in increased water loss through evaporation from the lungs.
HIGH ALTITUDE EXPOSURE: there is a marked increase in the loss of water through breathing in colder air, and fluid loss is further increased by the greater

respiratory efforts required to breathe at higher altitudes.
BURNS: burning destroys the outermost layers of the skin and any barrier to water diffusion, which then increases dramatically.
ILLNESS: water loss is increased if the victim suffers vomiting or diarrhoea.
Dehydration itself is a life-threatening condition. While not dangerous in the early stages, it can dull your mind and therefore reduce your overall will to survive. You must see thirst and dehydration as serious conditions. The symptoms are a loss of appetite, lethargy, impatience, doziness, emotional instability, slurred speech and a failure to be mentally coherent. The treatment is straightforward: replace lost fluids by drinking water. However, drink warm water first – the system will accept it more easily than cold water.

ROYAL MARINES TIPS

GUIDELINES FOR DEALING WITH DEHYDRATION

Britain's arctic warriors are experts in dealing with cold weather ailments, including dehydration, which can be a major problem in cold climates.

- Keep casualty warm.
- Loosen clothing to allow good circulation.
- Give patient liquids and salt gradually: one teaspoon of salt in two litres (0.44 gallons) of water.
- Allow patient plenty of rest.

MINIMISING WATER LOSS

Because water loss can bring on dehydration if the survivor does not have access to a regular supply of water, measures should be taken to reduce water loss.

All physical activity should be reduced to a minimum. Perform all tasks slowly to lessen the expenditure of energy, and have regular rest periods. In hot climates carry out essential activities at night or during the cooler parts of the day. In addition, keep clothing on to reduce fluid loss. There is a temptation to take off clothing in hot climates. Don't! Sweat in clothing cools the air trapped between the clothing and the skin, resulting in a decrease in the overall activity of the sweat glands and thus a reduction in water loss.

Wear light-coloured clothing in hot weather: it reflects the sun's rays and keeps any increase in body temperature to a minimum.

CARRYING WATER

Water bottles should be an essential item of any backpacker's kit. Plastic and aluminium models are available, though remember aluminium ones can be placed in a fire to heat the water if required.

BRITISH SAS TIPS

MINIMISING WATER LOSS

The soldiers of the British Special Air Service, who often have to survive behind enemy lines with few supplies, know how to cut down on fluid loss.

- Rest as much as possible.
- Avoid smoking and drinking alcohol; the latter uses fluids from the vital organs to break it down, and smoking increases thirst.
- Stay in the shade.
- Avoid lying on hot or heated ground and surfaces.
- Eat as little as possible: the body uses fluids to break down food; this can increase dehydration.
- Do not talk, and breathe through the nose not the mouth.

FINDING WATER

Though the location of water sources will be dealt with in the chapters concerning survival in specific types of terrain, all survivors should be aware of the water sources available to them and the methods of producing water.

Indicators of water sources include the following:

SWARMING INSECTS: look out especially for bees and columns of ants.

BIRDS: may gather around water, though note that water birds can travel for long distances without water and their presence may not indicate a water source. Birds of prey obtain liquids from their kills and are not indicators of a water source.

AN ABUNDANCE OF VEGETATION OF MANY VARIETIES: often this indicates that the plants are drinking from water that is near the surface.

ANIMALS: grazing animals need water at dusk and dawn, though meat eaters get liquids from their prey and may not indicate a local water source.

LARGE CLUMPS OF LUSH GRASS

ANIMAL TRACKS: often lead to water.

SPRINGS AND SEEPAGES IN ROCKY TERRAIN: limestone and lava rocks will have more and larger springs than other rocks. Lava rocks contain millions of bubble holes, through which water may seep.

CRACKS IN ROCK WITH BIRD DUNG OUTSIDE: may indicate a water source that can be reached by a straw.

VALLEY FLOORS: dig along their sloping sides to find water.

METHODS OF PRODUCING WATER

Solar still (Diagram 5) An excellent way of procuring water. Dig a hole 0.9m (3ft) across and 0.6m (2ft) deep, dig a sump in the middle of the hole and put a

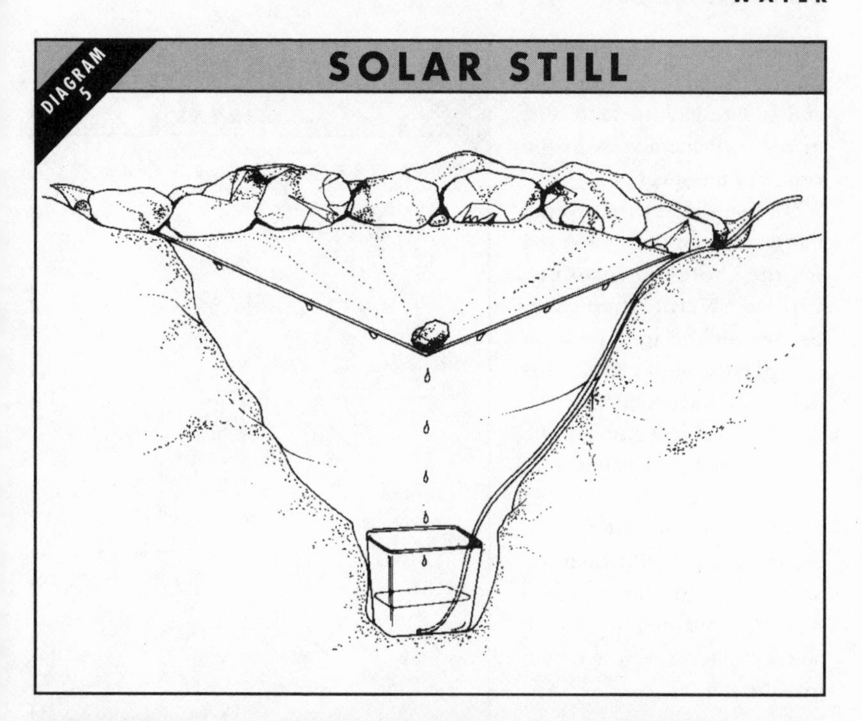
DIAGRAM 5
SOLAR STILL

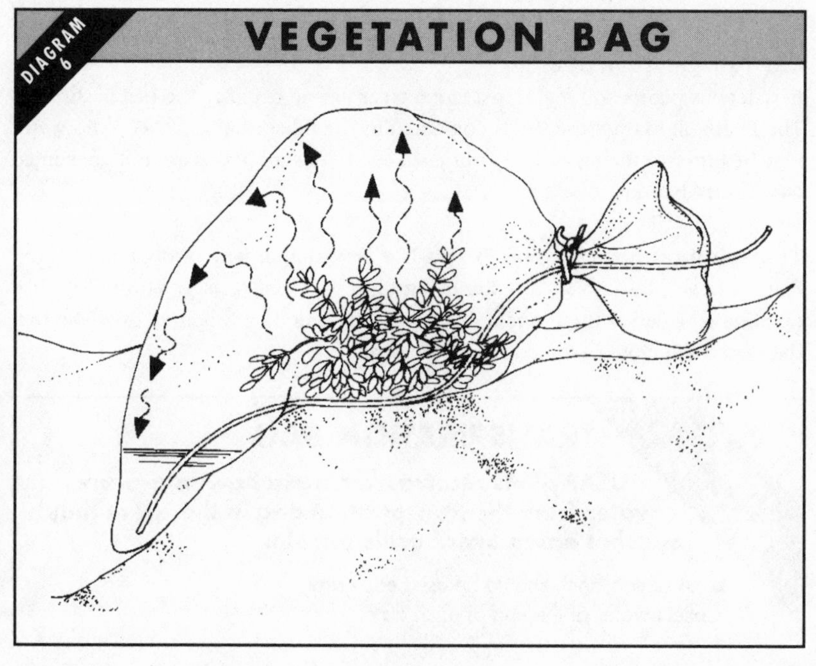
DIAGRAM 6
VEGETATION BAG

container in it, then place a plastic sheet over the hole and secure it with sand, dirt or rocks. Place a rock in the centre of the sheet.

The sun raises the overall temperature of the air and soil in the hole to produce vapour. Water then condenses on the underside of the plastic sheet and runs down into the container.

The solar still can also be used to distil pure water from contaminated water or sea water. Dig a small trench around 25cm (10in) from the still and pour the polluted water/sea water into it. The soil will filter it as it is drawn into the still.

Vegetation bag (Diagram 6) Cut foliage from trees or herbaceous plants and seal it in a large, clear plastic bag. Lay the bag in the sun. The heat will extract the fluids contained in the foliage. *WARNING* – the water may be bitter to the taste. Carry out taste test to ensure toxins are not consumed (see Food Chapter).

Transpiration bag (Diagram 7) This is a very simple way of obtaining water. Place a large plastic bag over the living limb of a tree or large shrub. The bag opening is sealed at the branch. The limb is then tied or weighted down so that the water will flow to the corner of the bag.

US AIR FORCE TIPS

TRANSPIRATION BAG

USAF pilots receive training in how to procure water from the transpiration bag in the event they are shot down over hostile terrain:

- A new branch should be used each day.
- Collect water at the end of each day.

Rain water A simple method of collecting rain water is to wrap a cloth around a slanted tree and ensure the bottom of the cloth drips into a container (Diagram 8).

METHODS OF FILTERING WATER

Remember, filtration does not purify water, it only removes the solid particles. Dig a hole along a water source and allow the soil to filter the water (Diagram 9). Alternatively, stretch layers of material across a tripod and fill each layer with grass, sand and charcoal (Diagram 10).

US SPECIAL FORCES TIPS

METHODS OF PURIFYING WATER

US Green Berets are taught these three simple ways of purifying water at the John F. Kennedy Special Warfare Center and School.

- Use water purification tablets (one in clear water, two in cloudy).
- Five drops of two per cent iodine in a container of clear water; 10 drops in cloudy or cold water (let it stand 30 minutes before drinking).
- Boil water for 10 minutes.

US AIR FORCE TIPS

IDENTIFYING CONTAMINATED WATER

Do not waste time in a survival situation by trying to purify contaminated water. Follow US Air Force training and avoid the following water sources:

- Those with strong odours, foam or bubbles in the water.
- Those with discolouration of the water.
- Lakes in deserts. Beware, they often have no outlets and are salt lakes.
- Those which lack healthy green plants around water source.

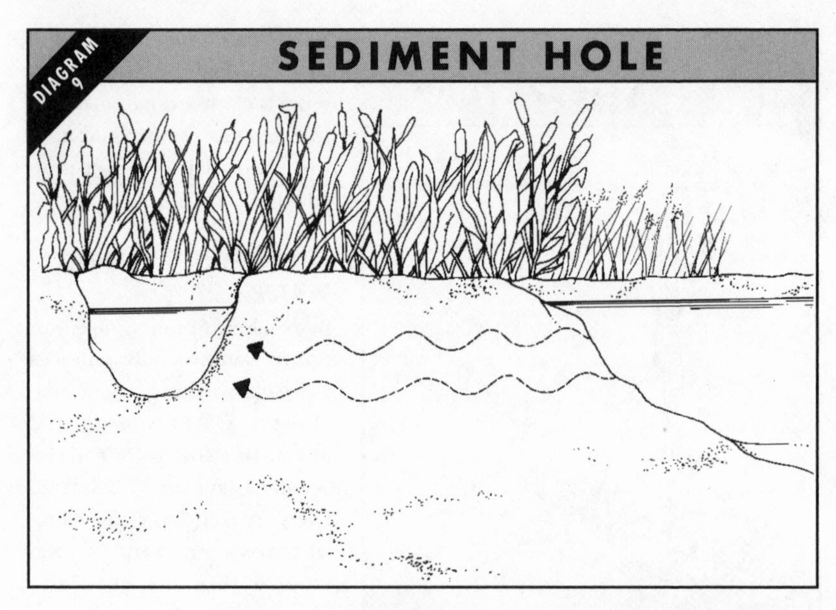

The charcoal removes bad smells and foreign objects from the water.

You can distil drinking water from sea water and urine. Place a tube into the top of a sealed fluid-filled container, ie sea water or urine, with the other end leading into a sealed empty container. Heat the container filled with water over a fire. This will make water vapour, which will travel through the tube and then cool, filling the empty container with drinkable water. If you are adrift in a raft on the ocean, this is an excellent way of obtaining drinking water, providing you have the means to heat the liquid (be careful when starting fires in dinghies).

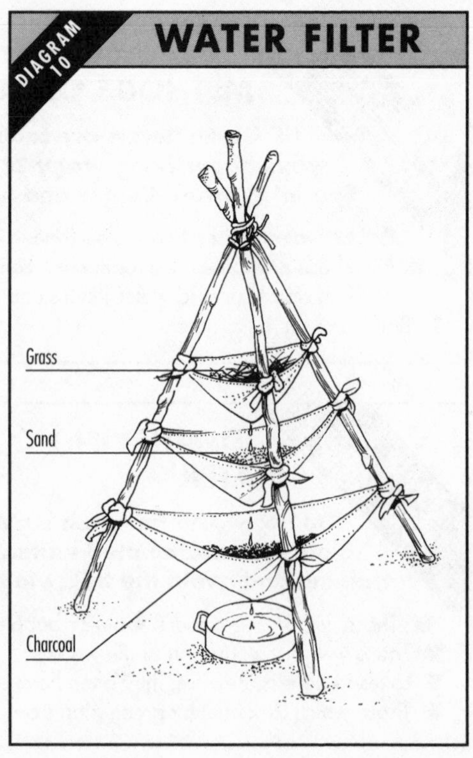

FIRE

As a survivor, you must learn how to make fire from the natural materials around you. Do not rely on matches or other commercial methods of making fire. This chapter explains the basic principles of how to start and maintain fires.

Fire is extremely important to the survivor both physically and psychologically. It is a great morale booster, keeps you warm, dries your clothes, boils water, can be used for signalling and for cooking food. It is therefore imperative that you know how to build, start and maintain a fire.

MAKING FIRE

The three ingredients of a successful fire are air, heat and fuel. The key to making a fire is to prepare all the stages of the materials and ensure all the ingredients are present. Be patient, and practise until you get it right.

The materials for a fire fall into three categories: tinder, kindling and fuel.

Tinder is any type of material that has a low flash point and is easily ignited. It usually consists of thin, bone-dry fibres. Tinder includes shredded bark from some trees and bushes, crushed fibres from dead plants, fine, dry wood shavings, straw and grasses, resinous sawdust, very fine pitch wood shavings, bird or rodent nest linings, seed down, charred cloth, cotton balls or lint, steel wool, dry powdered sap from pine trees, paper and foam rubber.

Get into the habit of always having tinder with you, and remember to carry it in a waterproof container.

Kindling has a higher combustible point and is added to the tinder. It is used to bring the burning temperature of the fire up to the point where less combustible fuel can be added to the fire.

Kindling includes dead, small, dry twigs, coniferous seed cones and needles, and wood that has been doused with flammable materials.

Fuel doesn't have to be dry, but moist wood will produce a lot of smoke. The best fuel sources are dry, dead wood and the insides of fallen trees and large branches (which may be dry even if the outside is wet). Green wood can be split and mixed with dry wood to be used as fuel. If there are no trees, twist dry grass into bunches, use dead cactus, dry peat moss or dried animal dung.

FIRE SITES

A site for your fire should be carefully selected. Remember that you want your fire to be a source of warmth, protection and cooking facilities.

BRITISH SAS TIPS

RULES FOR FIRE SITES

It is important to have a good spot for your fire. The SAS has many years experience of building survival fires in all types of terrain. It advises:

- Choose a sheltered site.
- Do not light a fire at the base of a stump or tree.
- Clear away all debris on the ground from a circle at least 2m (6ft) cross until you reveal bare earth.
- If ground is wet or covered with snow, build the fire on a platform constructed from green logs covered with a layer of earth or stones.
- In strong winds dig a trench and light a fire in it.
- In windy conditions encircle your fire with rocks.

If you have to build a fire in deep snow or marshy ground, build a temple fire. This consists of a platform of green logs with earth on top which is raised above the ground by four uprights at each corner, which have cross-pieces in their forks for the platform to rest on.

FIRE REFLECTORS

If you can, build a fire reflector (basically a wall) out of logs or rocks. It will direct or reflect the heat where you want it and will reduce the amount of wind blowing into the fire. You can use reflectors to direct heat into your sleeping shelter.

Do not build a fire up against a rock; rather, position it so you can sit between the rock and the fire. For even greater warmth, build a reflector beyond the fire that directs warmth back towards you (the rock will absorb warmth and keep your back warm).

DANGER

Do not place wet or porous rocks and stones near fires – they can explode when heated. Do not use slates and soft rocks, or any that crack, sound hollow or flake. Test all rocks by banging them together. If they contain moisture they will expand faster than the rock when heated and may explode, with potentially lethal effects.

STARTING FIRES WITHOUT MATCHES

Survivors must know how to start a fire without matches in a long-term survival situation. There are a number of easy ways to make fire without commercial matches, four of which are shown in Diagram 11. When trying to start a fire, remember to do so out of the wind or with your back to the wind.

Flint and steel (11A) Hold the flint and steel above the tinder. Strike the flint with the edge of the steel in a downward glance. The sparks must be fanned on the tinder and then further blown or fanned to produce a coal and subsequent flame.

> **WARNING**
>
> ***Keep sparks and flames away from the battery because explosive hydrogen gas is produced, which can cause serious injury.***

Battery (11B) If you have access to a battery, connect the end of one piece of insulated wire to the positive post and the end of another piece of insulated wire to the negative post. Touch the two remaining ends to the ends of a piece of non-insulated wire. The non-insulated wire will begin to glow and get hot, and can be used to ignite kindling. Remember to move the battery away once you have the fire going.

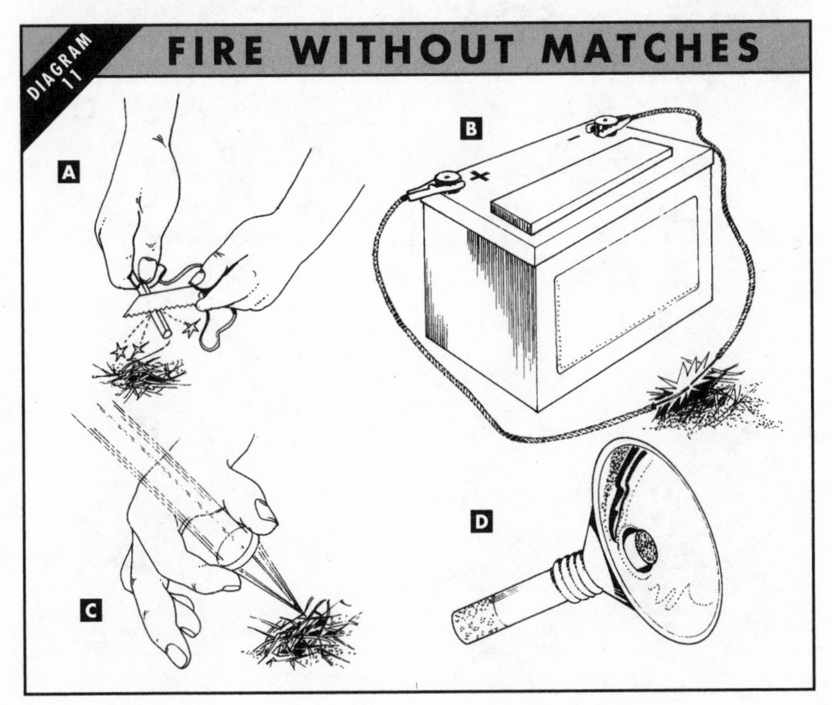

Burning glass (11C) Concentrate rays of the sun on tinder using a magnifying glass, a camera lens, the lens of a flashlight that magnifies, or even a convex piece of bottle glass.

Flashlamp reflector (11D) Place tinder in the centre of reflector where the bulb is usually located. Push it up from the back of the hole until the hottest light is concentrated on the end and smoke results. If available, a cigarette can be used as tinder for this method.

Bow and Drill (Diagram 12) An ancient method of making fire, and a useful one for the survivor to know.

Make a spindle out of straight hardwood, the spindle to be around 30-45cm (12-18in) long and 1.9cm (0.75in) in diameter. Round one end and work the other into a blunt point. The round end goes into the socket (which is made from hardwood and which can be held comfortably - put grease or soap in the hole to prevent friction).

The bow should be made from a branch around 0.9m (3ft) long and 2.54cm (1in) in diameter. Tie a piece of suspension line or leather thong to both ends so it has the tension of a bow (A).

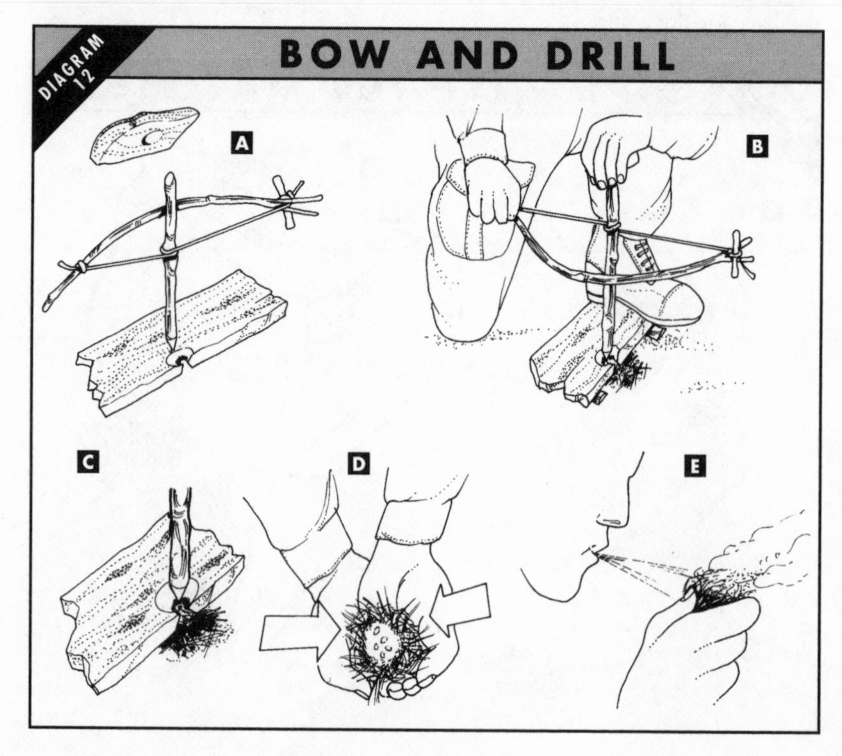

The fireboard is made from softwood and is around 30cm (12in) long and 1.9cm (0.75in) thick and 7.5-15cm (3-6in) wide. Carve a small hollow in it then make a V-shaped cut in from the edge of the board. It should extend into the centre of the hollow, where the spindle will make the hollow deeper. The object of the 'V' is to create an angle that cuts off the edge of the spindle as it gets hot and turns to charcoal dust.

While kneeling on one knee, place the other foot on the fireboard and place tinder just beneath the V-cut. Rest the board on two sticks to create the space (this allows air into the tinder).

Twist the bow string once around the spindle and place spindle upright into the hollow. Then press socket down on the spindle and fireboard. Spin the spindle with long, even strokes of the bow until smoke is produced (B). The spinning should become faster until thick smoke is produced. By this stage a hot powder that can be blown into a glowing ember has been produced (C).

US SPECIAL FORCES TIPS

FIRECRAFT TIPS

Use these very simple rules, employed by America's elite soldiers, for making and siting fires in the wild.

- Reserve your matches for starting properly prepared fires, not for lighting cigarettes or improperly prepared fires.
- Always try to carry dry tinder in a waterproof container.
- In the arctic, a platform will be needed to prevent fire from melting down through deep snow and putting it out.
- You will also need a platform if you start a fire in an area of peat or humus to stop it spreading – a smouldering peat fire can burn for years.
- In woods, clear away ground debris to prevent fire from spreading.

The bow and spindle can be removed and tinder placed next to the glowing ember (D). Roll tinder around the burning ember and blow to burn the tinder (E). The burning tinder is then placed in a waiting fire containing more tinder and small kindling.

TYPES OF FIRE

There are different types of fires, all used for specific purposes. The most common used by survivors are shown in Diagram 13. You should learn how to site and build them all.

Safety night fire (A) This fire enables you to stay close to the fire while sleeping without the danger of burning logs rolling on you. Place two large, green logs against the fire, ensuring that as the fire burns it will be pushed away from you and your shelter. Make sure the fire has as few air spaces as possible to ensure it burns throughout the night. Note the position of the heat reflector.

Long fire (B) This fire begins as a trench, which is dug to take advantage of the wind. However, it can also be constructed above ground by using two parallel green logs to hold the coals together. Make sure the logs are at least 15cm (6in) wide (although the thicker the better) and positioned so that cooking utensils can be rested on them. Two 2.5cm (1in) diameter sticks can be placed under them to allow the fire to receive more air.

'T' fire (C) This fire is good for cooking. The fire is maintained in the top part of the 'T', which provides coals for cooking in the bottom part of the 'T'.

Tepee (D) A good fire for both cooking and heat (remember to have ample supplies of fuel). Place tinder in middle of fire site and push a stick into the ground, slanting over the tinder. Lean a circle of kindling sticks against the slanting stick, with an opening towards the wind for draft. Light the fire with your back to the wind and feed the fire from the downwind side.

Star fire (E) Used to conserve fuel or if you want a small fire. The fire is in the centre of the 'wheel'; the logs are pushed in according to needs. They can be drawn apart if cooking over embers is required. Hardwood is recommended for this type of fire.

Keyhole Fire (F) Dig a hole in the ground in the shape of a key, taking advantage of the wind. This fire does the same job as the long fire.

Pyramid fire (G) Similar to the log cabin fire, except there are layers of fuel instead of a hollow framework. This fire burns for a long time and can be used as an overnight fire.

Log cabin (H) This fire gives off great heat and light because of the amount of oxygen which enters it. As such, it can be used for cooking and signalling.

CARRYING FIRE

Carrying fire is an effective way of preserving your fire-starting materials and saving you the bother of starting a fire when you decide to make camp for the night. Carrying fire was used by prehistoric man and is still used by primitive peoples throughout the world. As with everything else concerning survival, you should practise making fire bundles and tubes before you actually need them, and remember to have more than one ready to use.

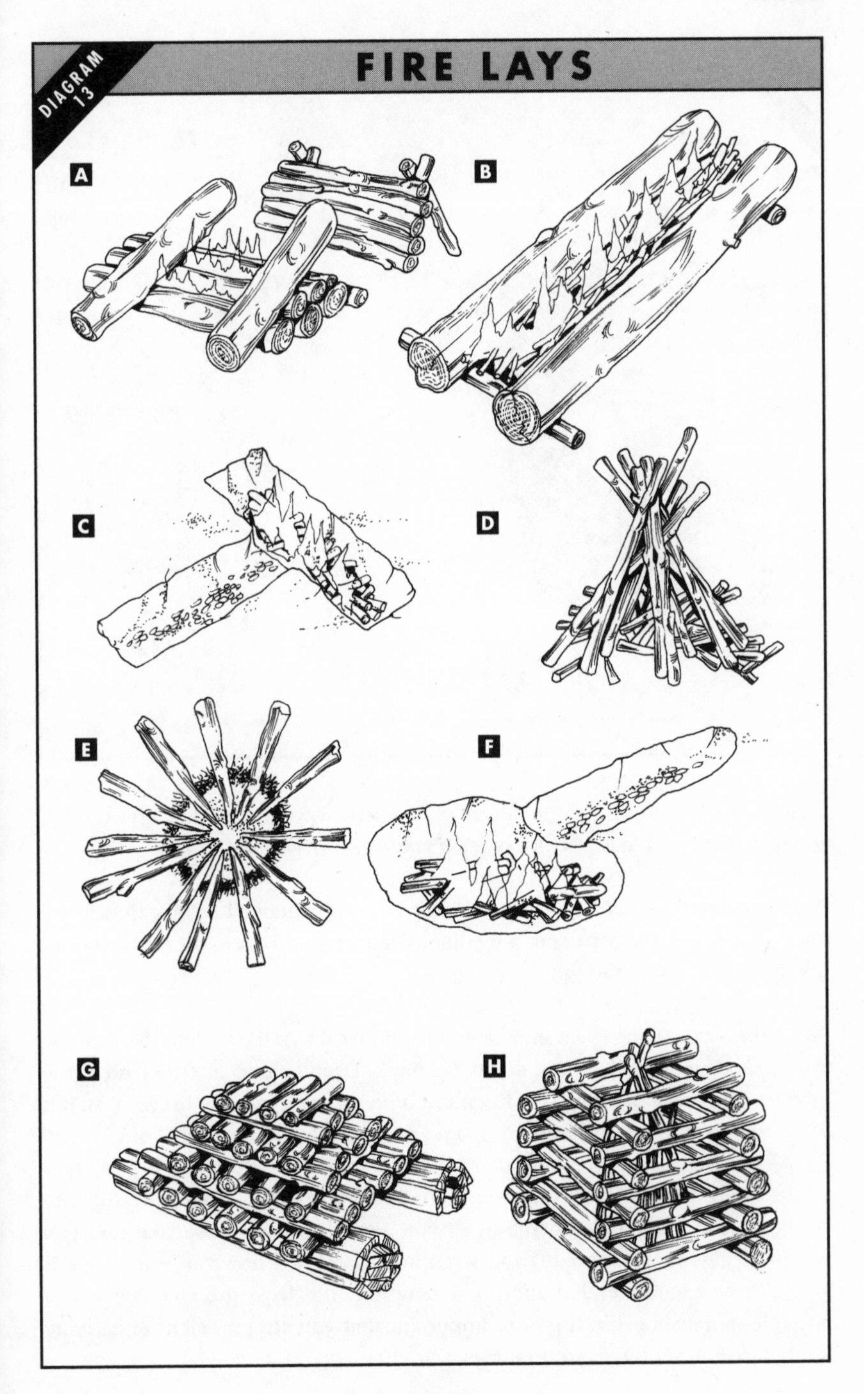
DIAGRAM 13
FIRE LAYS
A
B
C
D
E
F
G
H

Diagram 14 shows two of the most effective ways of carrying fire, though you may only be able to use fire tubes if you do not have a tin can.

Fire bundles (A) Place a number of hot coals, surrounded by dry tinder and then damp grass and leaves, in a medium-sized tin can. Ensure there are ventilation holes pierced in the can.

Fire tube Take a long sheet of bark and lay tinder down the middle (B). Roll the tube and secure it with ties all down its length. Drop embers into the end of the tube to start it smouldering (C). Keep the tube pointing into the breeze (D); if it catches fire spit on it or stamp on it to regain control.

There other ways of carrying fire, including transporting a burning log and swinging it to keep it alight (though if you are going to do this make sure you are physically strong enough!) and wrapping a coal in a fireproof leaf. There is much to be recommended about carrying fire, not least the fact that it is a portable morale-booster. It is very important that you are proficient in carrying fire before you set off on your journey.

FOOD

No matter where you are, there are many types of plant and animal foods available. The survivor must learn where to find them, how to recognise them and how to collect or trap them, and must also know what is poisonous and dangerous.

YOUR REQUIREMENTS

The average energy requirements are up to 3000 calories a day for a man and up to 2000 calories for a woman. However, in a survival situation, where you undertake more strenuous activity, you will need more: 3-5000 calories a day in warm weather and 4-6000 calories a day in cold climates.

You must try to eat a balanced diet to keep your body and mind working properly. This means ensuring that you eat each of the major food constituents on a daily basis. These constituents are:

PROTEIN: essential for growth and repair of tissue. Also an energy source when inadequate amounts of carbohydrates and fats are present. Protein is broken down into amino acids, which are essential for muscular growth and repair. Protein is found in cheese, milk, cereal grains, fish, meat and poultry.

CARBOHYDRATES: very simple molecules that are easily digested. They are the body's main source of energy. For the survivor, they should constitute up to half of the daily calorific intake. Carbohydrates can be obtained from fruit, vegetables, chocolate, milk and cereals.

FATS: the most concentrated energy source. Fats are utilised by the body when its carbohydrate stores have been depleted. Fats can be obtained from butter, cheese, oils, nuts, egg yolks, margarine and animal fats.

VITAMINS: regulate the body's vital functions. There are many kinds, but most can be obtained through maintaining a balanced diet.

MINERALS: regulate the body's functions and form vital constituents of teeth and bones. Like vitamins, a well-balanced diet should ensure an adequate supply of minerals.

COOKING

The main types of cooking are boiling, frying, parching, baking, steaming and roasting, and you can employ them all in the wild.

BRITISH SAS TIPS

SURVIVAL COOKING

SAS soldiers are expert at turning even the most unlikely looking and seemingly inedible creature into a tasty and nutritious meal.

MEAT: cut into small cubes and boil. Treat pork in warm climates with caution, wild pig is infested with worms and liver fluke, and venison is also prone to worms.
FISH: usually germ-free if caught in fresh water; best stewed or wrapped in leaves and baked.
BIRDS: boil all carrion; young birds can be roasted.
REPTILES: gut and cook in their skins in hot embers; when skin splits remove and boil; cut off snake heads before cooking – some have venom glands in their heads. Skin frogs (many have poisonous skins) and roast on a stick.
TURTLES AND TORTOISES: boil until shell comes off; cut up meat and cook until tender.
SHELLFISH: boil crabs, lobsters, shrimps, crayfish and prawns to remove harmful organisms; all seafood spoils quickly – cook as soon as possible.
INSECTS AND WORMS: can be boiled, or dried on hot rocks, crushed and then ground into a powder to add to soups and stews.

Boiling Food can be boiled in a metal container, in a rock with a hole in it (not ones with a high moisture content - they can explode and inflict serious injuries) and a hollowed-out piece of wood (hang wood over the fire and add hot rocks to the water and food, replace as they cool with hot ones until food cooks). Food can also be boiled in coconut shells, sea shells, turtle shells and half sections of bamboo.

Frying Put a flat piece of rock on a fire; when it is hot fry food on it.

Parching Works well with nuts and grains. Place in a container or on a rock and heat slowly until they are scorched.

Baking Improvise an oven by using a pit under a fire, a closed container or a wrapping of leaves or clay. Another method of baking is to line a pit with moisture-free stones and build a fire in the pit. As the fire burns down, scrape the coals back, put the covered container in, cover with a layer of coals and a thin layer of dirt. The food will then bake.

Steaming Also done in a pit. Wrap the food in large leaves or moss. Place one layer of food in a coal-lined pit. Add another layer of leaves and moss and continue alternating layers of wrapped food and leaves or moss until the pit is almost full. Push a stick through the layers of food and leaves or moss then seal the pit with dirt.

Roasting Done with a skewer or spit over an open fire. Good for cooking whole fowls or small animals.

FOOD FROM PLANTS

There are thousands of edible plants available throughout the world. The survivor should carefully examine the area he or she is in for edible plant species. Some of the most common are illustrated in Diagram 15.

When selecting unknown plants to eat, you MUST carry out the taste test to see if they are safe to eat (see box). It is important to test *ALL* parts of the plant - many plants have only one or more edible parts.

> **WARNING**
>
> ***While many fungi are edible, others are deadly poisonous. Although fungi are discussed later in the chapter, you may be well advised to stay clear of fungi altogether as a food source.***

Edible underground parts of plants Tubers, usually found below ground; are rich in starch and should be roasted or boiled. Plants with edible tubers include arrowhead (K), tara, yam, cat's-tail (L) chufa and sweet potato.

ROOTS AND ROOTSTALKS are rich in starch. Plants with edible rootstalks include baobab, goa bean, water plantain, bracken, reindeer moss, cow parsnip

> US AIR FORCE TIPS
>
> **CAN YOU EAT THIS PLANT?**
>
> **Use these simple US Air Force guidelines, drawn up for downed pilots, when selecting plants for possible consumption in the wild. They will serve you well in a survival situation.**
>
> - Avoid plants with umbrella-shaped flowers, though carrots, celery and parsley (all edible) are members of this family.
> - Avoid all legumes (beans and peas); they absorb minerals from the soil and cause digestive problems.
> - If in doubt, avoid all bulbs.
> - Avoid all white and yellow berries – they are poisonous; half of all red berries are poisonous; blue or black berries are generally safe to eat.
> - Aggregated fruits and berries are edible.
> - Single fruits on a stem are considered safe to eat.
> - A milky sap indicates a poisonous plant.
> - Plants that are irritants to the skin should not be eaten.
> - Plants that grow in water or moist soil are often very tasty.

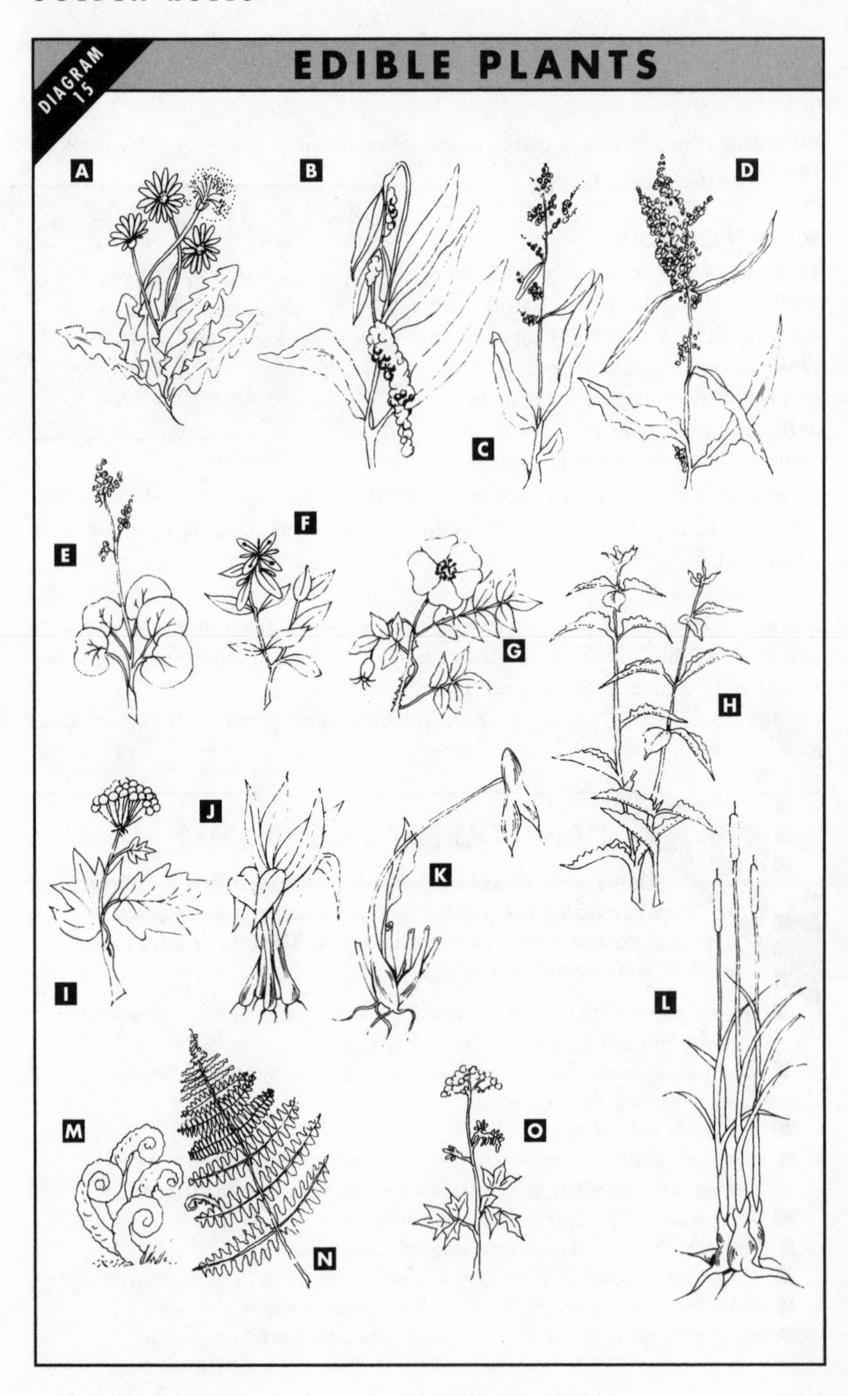
DIAGRAM 15
EDIBLE PLANTS
A
B
C
D
E
F
G
H
I
J
K
L
M
N
O

(I), wild calla, rock tripe, canna lily, cattail, chicory, horseradish, tree fern (M & N), lotus lily, Angelica (O) and water lily.
BULBS can be poisonous: e.g. death camas, which has white or yellow flowers. However, others are edible: e.g. wild lily, wild tulip, wild onion (J), blue camas and tiger lily.
EDIBLE SHOOTS grow in much the same way as asparagus. Many can be eaten raw, but they are better boiled. Edible shoots include purslane, reindeer moss, bamboo, fishtail palm, goa bean, bracken, rattan, wild rhubarb, cattail, sago palm, rock tripe, papaya, sugar cane and lotus lily.

Plants with edible leaves are perhaps the most numerous of all edible plants and include dandelion (A), fireweed (F), dock (B-D), mountain sorrel (E) and nettle (H). In addition, the young tender leaves of nearly all non-poisonous plants are edible.

Some plants with edible leaves have an edible pith in the centre of the stem. Examples include buri, fishtail, sago, coconut, rattan and sugar cane.

US ARMY TIPS

TASTE TEST

Use this simple US Army test for establishing whether a plant is safe to eat. IT CANNOT BE APPLIED TO FUNGI.

- Test only one part of the plant at a time.
- Break plant into its base constituents: leaves, stem, roots, etc.
- Smell the plant for strong or acid odours.
- Do not eat for eight hours before starting the test.
- During this period put a sample of the plant on the inside of your elbow or wrist. 15 minutes is enough time to allow for a reaction.
- During the test period take nothing orally except pure water and the plant to be tested.
- Select a small portion of the component.
- Before putting it in your mouth, put the plant piece on the outer surface of the lip to test for burning or itching.
- If after three minutes there is no reaction, place it on your tongue; hold for 15 minutes.
- If there is no reaction, chew a piece thoroughly and hold it in your mouth for 15 minutes. *DO NOT SWALLOW.*
- If there is no irritation whatsoever during this time, swallow the food.
- Wait eight hours. If any ill effects occur induce vomiting and drink plenty of water.
- If no bad effects occur, eat half a cup of the same plant prepared the same way. Wait another eight hours; if no ill effects are suffered the plant as prepared is safe to eat.

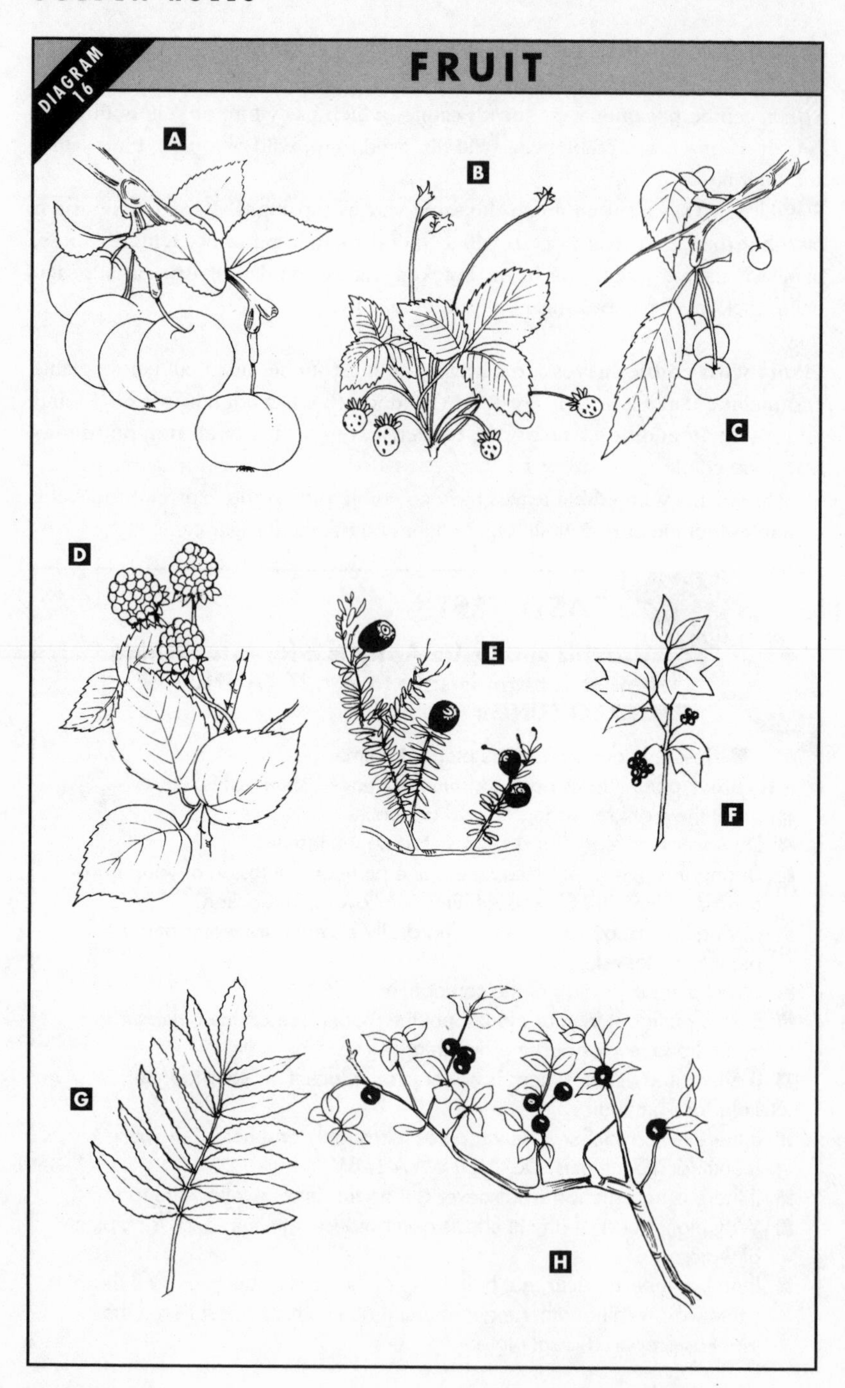
DIAGRAM 16
FRUIT
A
B
C
D
E
F
G
H

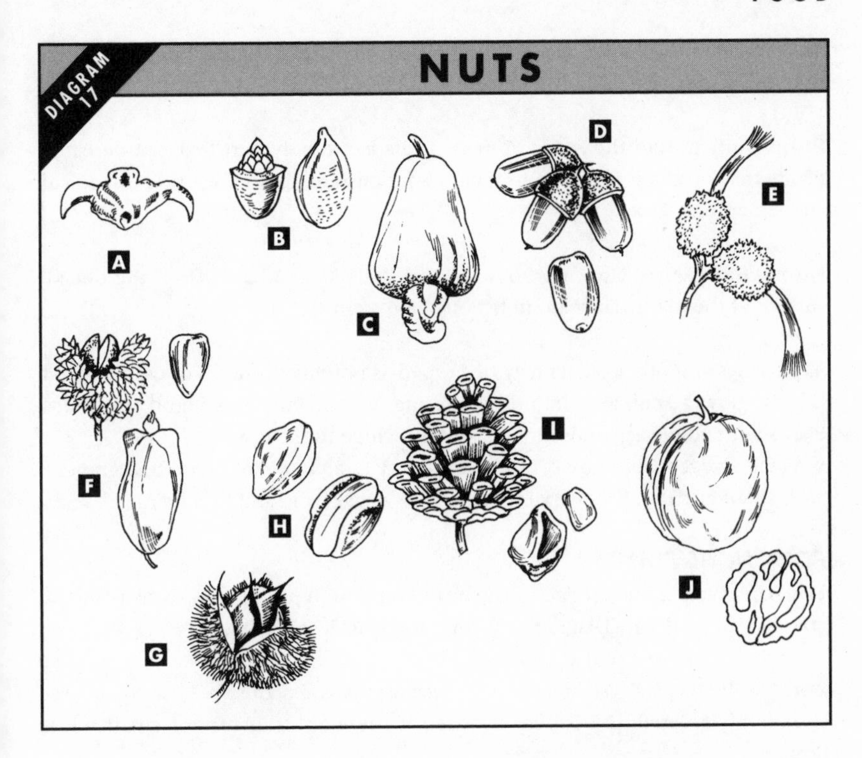

The inner barks of a tree (the layer next to the wood) may be eaten raw. Avoid the outer bark, which contains large amounts of bitter tannin.

Flower parts Edible flowers include abal, wild rose (G), colocynth, papaya, banana, horseradish, wild caper and luffa sponge. Pollen has the appearance of yellow dust and is high in food value.

Fruits (Diagram 16) There are many kinds of edible fruits, both of the sweet and non-sweet (vegetable) type. Sweet fruits include crab apples (A), wild strawberry (B), wild cherry (C), blackberry (D), crowberry (E) and cranberry (F). Vegetables include breadfruit, horseradish, rowan (G & H) and wild caper.

Seeds and grains The grains of all cereals and other grasses are good sources of plant protein. They can be ground up and mixed with water to make porridge. Plants with edible seeds and grains include amaranth, Italian millet, rice, bamboo, nipa palm, tamarind, screw pine, water lily and purslane.

Nuts (Diagram 17) are a good source of protein. Most can be eaten raw, though some, such as acorns (D), are better when cooked. Plants with edible nuts include almond (B & H), water chestnut (A), beechnut (F), oak, pine (I), chestnut (G), cashew (C), hazelnut (E) and walnut.

Pulps Pulp around the seeds of many fruits is the only part that can be eaten. Plants that produce edible pulp include the custard apple, inga pod, breadfruit and tamarind.

Gums and resins They are both sap that collects and hardens on the outside surface of the plant. They are nutritious food sources.

Saps Vines and other plants may be tapped as potential sources of usable liquid. Cut the flower stalk and drain the fluid into a container. It is highly nutritious. Plants with edible sap and drinking water include the following:
WATER: sweet acacia, colocynth, agave, saxual, rattan palm, cactus and grape.
SAP: coconut palm, fishtail palm, sago palm, sugar palm and buri palm.

POISONOUS PLANTS

You must have a knowledge of the most common types of poisonous plants so you can avoid them (Diagram 18). In particular, learn to identify hemlock and water hemlock, two of the deadliest.
WATER HEMLOCK (G) Appearance: purple-streaked stems, hollow-chambered rootstalk (I), small 2-3 lobed, toothed leaflets and clusters of small white flowers.
Location: always found near water throughout the world.
Smell: unpleasant odour. *DEADLY*
HEMLOCK Appearance: up to 2m (6ft) high, multi-branched. Hollow, purple-spotted stems, coarsely toothed leaves, lighter below. Dense clusters of tiny white flowers and white roots.
Location: grassy waste places throughout the world.
Smell: unpleasant smell. *DEADLY*
POISON IVY Appearance: three-part variable leaves, greenish flowers and white berries (this is a contact poison).
Location: wooded areas of North America.
BANEBERRY (A) Appearance: leaves made up of several toothed leaflets. Small, white flowers clustered at the end of a stem and white or black berries.
Location: mostly in woods.
POISON SUMAS Appearance: hairless, oval leaflets in opposite pairs, dark-spotted smooth bark and clusters of white berries (this is a contact poison).
Location: swamplands of southeastern North America.
DEATH CAMAS (C) Appearance: long, strap-like leaves and loose clusters of greenish-white, six-part flowers.
Location: grassy, rocky and lightly wooded areas of North America. *DEADLY*

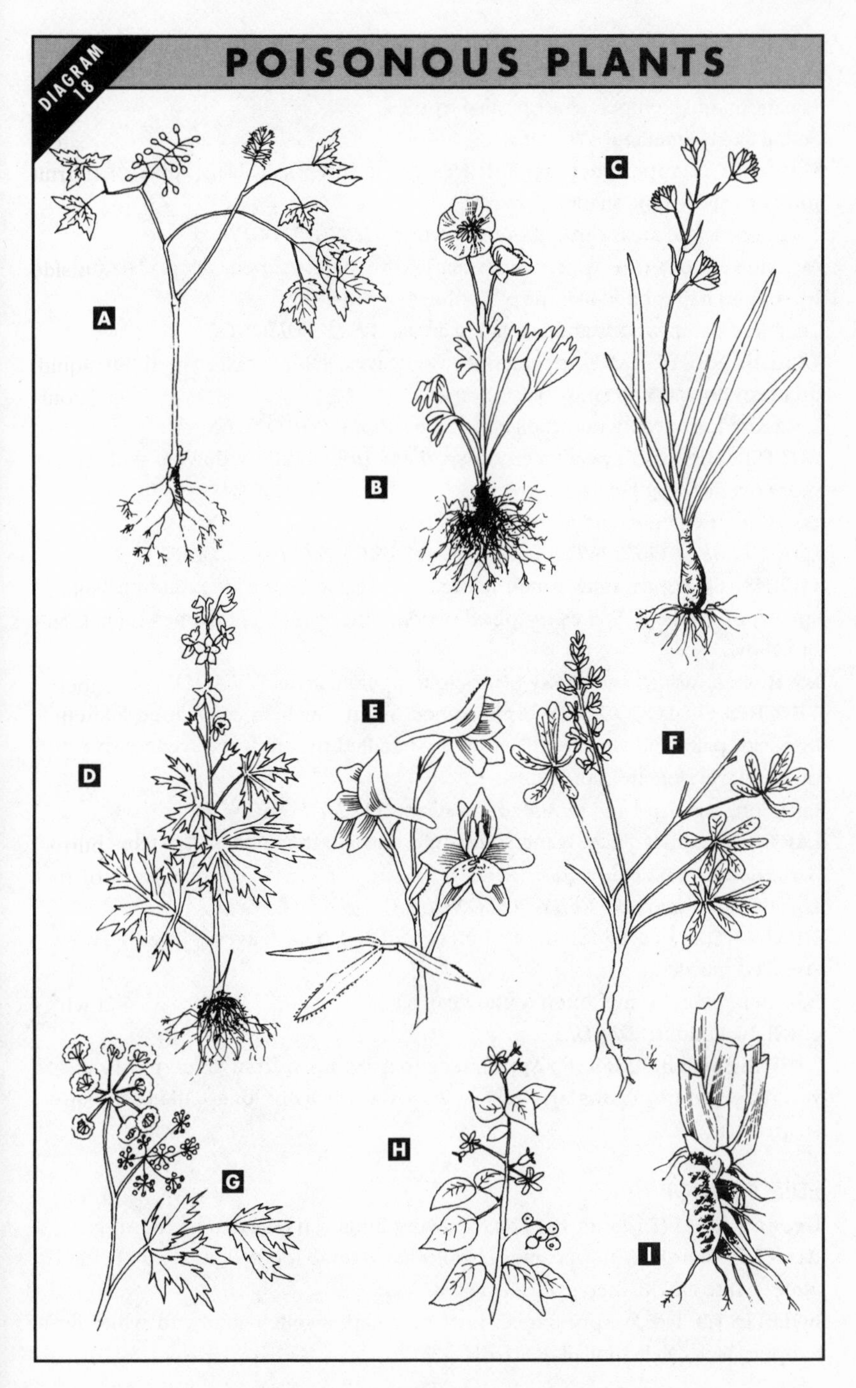
DIAGRAM 18
POISONOUS PLANTS
A
B
C
D
E
F
G
H
I

THORN-APPLE OR JIMSON WEED Appearance: jagged-toothed oval leaves and single large, trumpet-shaped white flower and spiny fruit.
Location: most temperate areas, also tropics.
Smell: sickly smelling. *DEADLY*
FOXGLOVES Appearance: basal leaves topped by a tall, leafy spike of purple, pink or yellow tube-shaped flowers.
Location: waste areas throughout the world. *HIGHLY TOXIC*
MONK'S-HOOD (D) Appearance: leafy with palm-shaped, deeply segmented leaves and hairy, hood-like purplish-blue or yellow flowers.
Location: damp woods and in shaded areas. *VERY POISONOUS*
DEADLY NIGHTSHADE Appearance: oval leaves, solitary bell-shaped purplish or greenish flowers and shiny black berries.
Location: European woodlands and scrub. *VERY POISONOUS*
BUTTERCUPS (B) Appearance: glossy, waxy bright yellow flowers with five or more overlapping petals.
Location: temperate and arctic areas.
CAN CAUSE SEVERE INFLAMMATION OF THE INTESTINAL TRACT
LUPINS (F) Appearance: small leaflets in a palm shape or radiating like the spokes of a wheel. Spikes of 'pea-flowers': blue, violet, sometimes pink, white or yellow.
Location: clearings and grassy places in temperate areas. *DEADLY*
VETCHES OR LOCOWEEDS Appearance: many small, spear-shaped leaflets in opposite pairs and showy spikes of five-petalled 'pea-flowers': yellowish-white, pink to lavender, and purplish.
Location: grassland and mountain meadows. *VERY POISONOUS*
LARKSPUR Appearance: leaflets radiating out like the spokes of a wheel. It has dark purple or blue flowers
Location: moist areas. *VERY POISONOUS*
HENBANE Appearance: sticky hairs, toothed oval leaves. Creamy flowers streaked purple.
Location: bare ground, often found near sea.
Smell: bad odour. *DEADLY*
NIGHTSHADE BERRIES (H) Appearance: berries ripen from green to black, red, yellow or white. Plants are bushy, leaves are usually long-stalked and spear shaped. *AVOID*

EDIBLE FUNGI

Ground fungi (Diagram 19) The following fungi can be eaten by the survivor:
GIANT PUFFBALL (F) Appearance: looks like a large football. Around 30cm (1ft) wide, white and leathery, yellowing with age.
BOLETES (D, I & J) Appearance: brownish cap, swollen stem and white flesh. Slippery Jack (A) is similar.

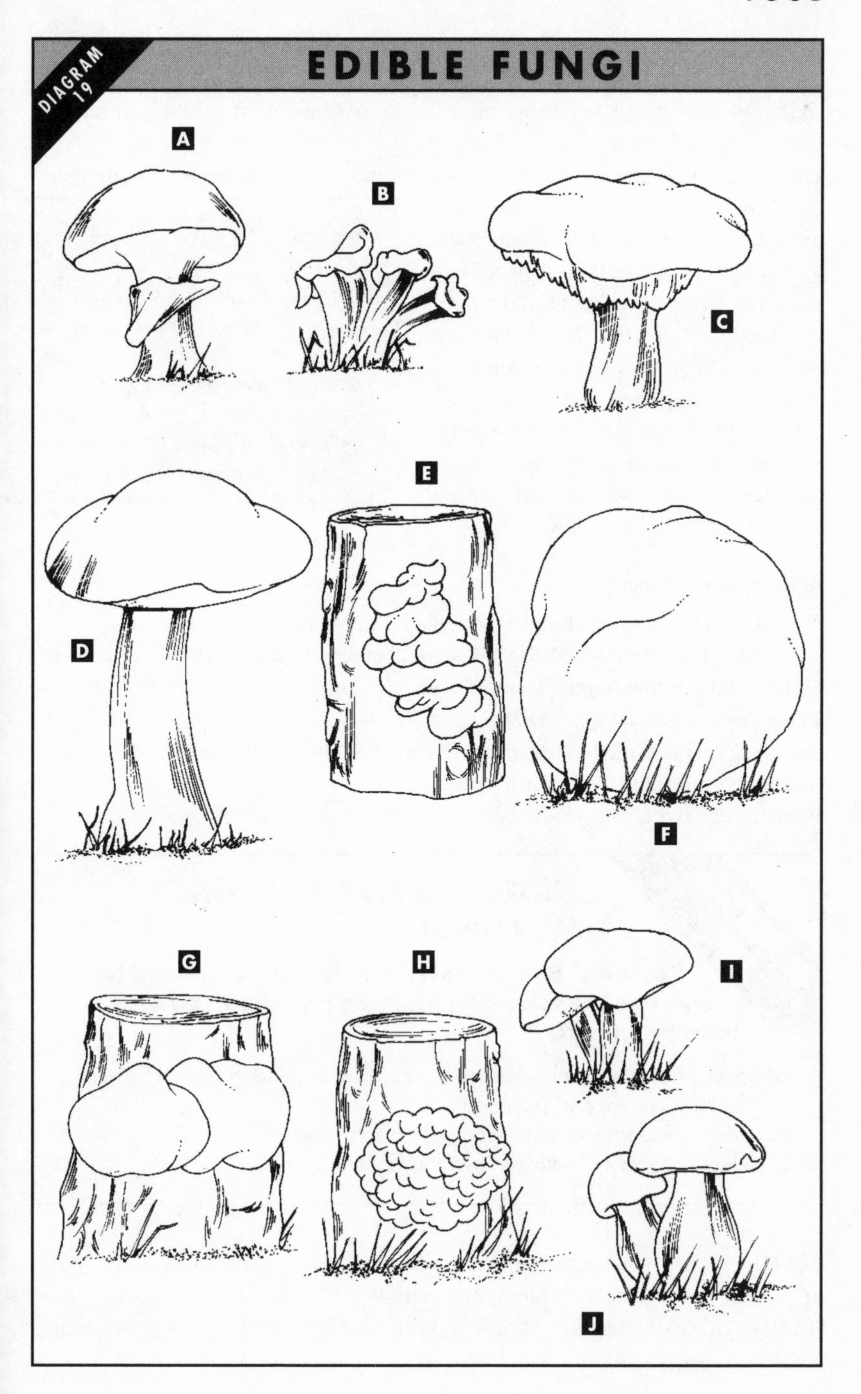
DIAGRAM 19
EDIBLE FUNGI
A
B
C
D
E
F
G
H
I
J

HORN OF PLENTY (B) Appearance: horn- or funnel-shaped. Has a rough, crinkly dark-brown cap and a smooth, tapering grey stem.
HEDGEHOG FUNGUS (C) Appearance: has spines instead of gills or pores.

Tree fungi (Diagram 19) The following tree fungi can be eaten by the survivor:
BEEFSTEAK FUNGUS (G) Appearance: reddish above, pinkish below and rough-textured, looks like a large tongue with blood-like juice. Tastes rough and bitter.
POLYPORUS SULPHUREUS Appearance: bright orange-yellow fading to yellowish-tan. Spongy, yellowish flesh.
BRAIN FUNGUS (H) Appearance: resembles a brain or coral.
CHICKEN-OF-THE-WOODS (E) Appearance: bright lemon or yellow.

WARNING

Fungi cannot be subjected to the taste test. Deadly fungi do not taste unpleasant and the symptoms of poisoning may not appear until several hours after eating. There is no antidote for fungal poisoning, so either positively identify the fungi you are going to eat or leave well alone.

POISONOUS FUNGI

The following poisonous fungus (See Diagram 20) MUST NOT be eaten:
YELLOW STAINING MUSHROOM Appearance: shows a yellow stain when bruised and is strongly yellow at the base.
Smell: smells of carbolic *AVOID*
DESTROYING ANGEL (A) Appearance: totally white, with a large volva, a scaly stem and a cap up to 12cm (5in) across.
Smell: smells sweet and sickly *DEADLY*

BRITISH SAS TIPS

GUIDELINES FOR SELECTING EDIBLE FUNGI

You must be extremely careful selecting fungi to eat. Use the following SAS tips for when you are collecting fungi.

- Avoid any fungi with white gills, a cup-like appendage at the base of the stem (volva) and stem rings.
- Avoid any fungi that are decomposing or wormy.
- Unless positively identified – avoid altogether.

DEATH CAP (B) Appearance: greenish-olive cap up to 12cm (5in) across, paler stem, large volva (C & D), white gills and flesh. *DEADLY*
FLY AGARIC (E) Appearance: bright red cap flecked with white, large volva (F-I)

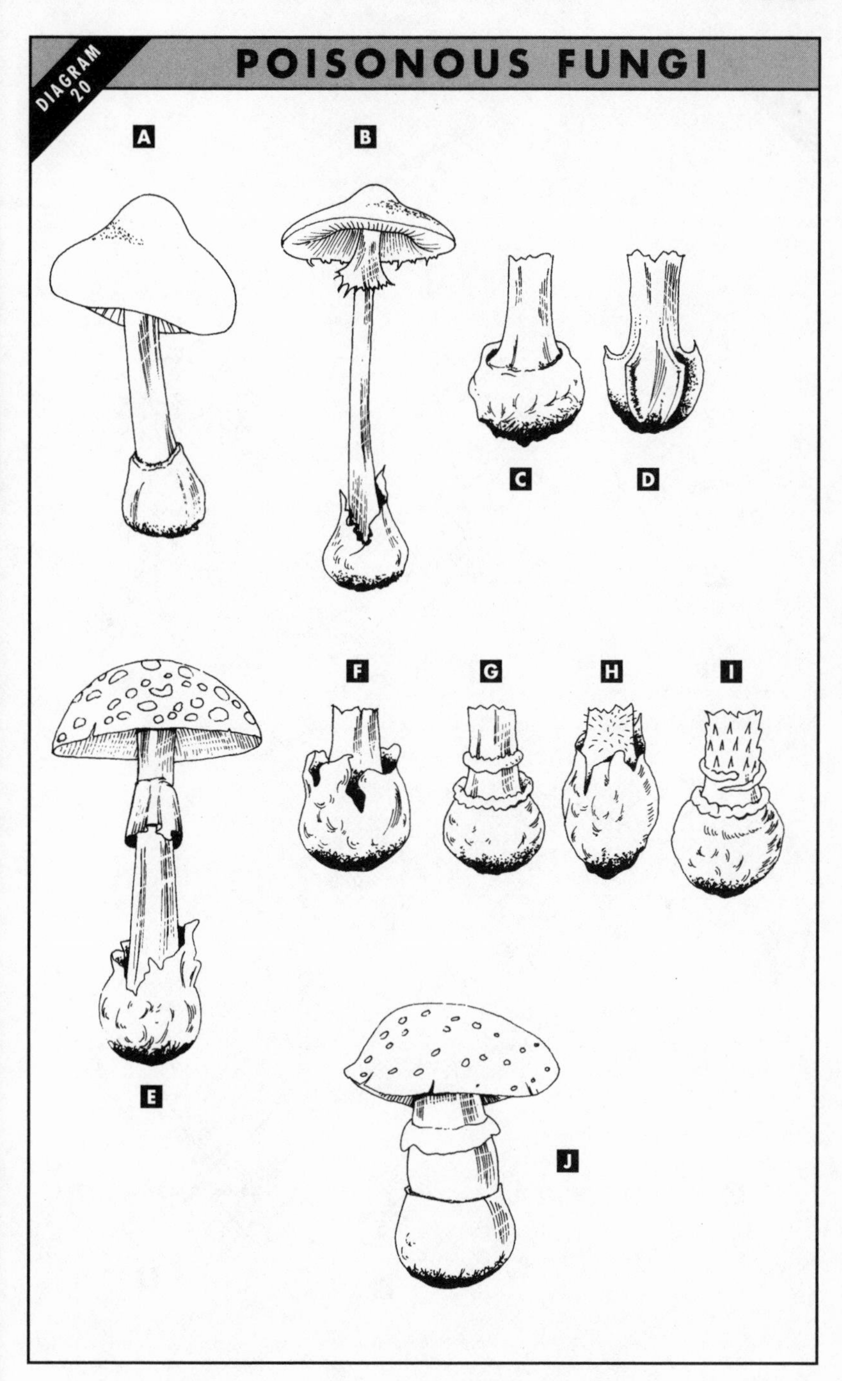
DIAGRAM 20
POISONOUS FUNGI
A
B
C
D
E
F
G
H
I
J

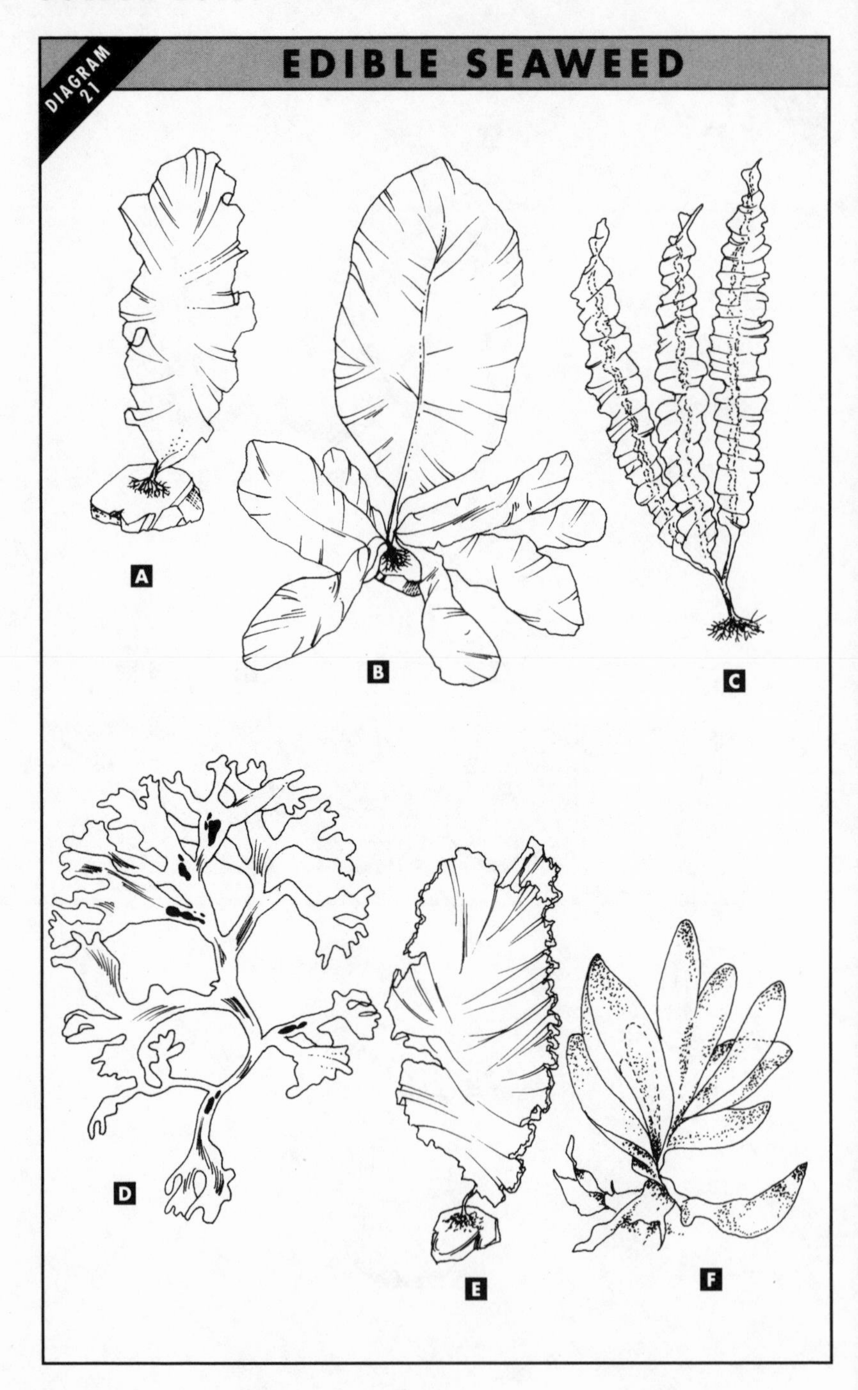
DIAGRAM 21
EDIBLE SEAWEED
A
B
C
D
E
F

up to 22cm (9in) across. *DEADLY*
PANTHER CAP (J) Appearance: brownish, white-flecked cap up to 8cm (3in) across, white gills and 2-3 hoop-like rings at base of stem. *DEADLY*
LEADEN ENTOLOMA Appearance: dull, greyish-white, deeply convex cap up to 15cm (6in) across. Yellowish gills turning salmon-pink and firm white flesh. Smell: bitter almond and radish. *DEADLY*

Seaweed (Diagram 21) Edible seaweed can be found in shallow waters anchored to the bottom of rocks, or can be found floating on the open seas. Sea lettuce (A) is a light green, kelp (B) is olive-green, sugarwrack (C) has long, flat yellow-brown fronds, Irish moss (D) has purplish to olive-green fronds, dulse (E) is purplish-red, and lavers (F) have red, purplish or brown fronds.

FOOD FROM ANIMALS

Remember one thing about acquiring food from animals: never expend more energy getting the food than you receive from it once you have caught and cooked it. You must become adept at hunting and trapping to use animals as a food source. Learn about the types of animals that inhabit the area you are in, their tracks (Diagram 22), habits and where they sleep.

The best animals for flavour and amount of meat are mature females, but all animals will provide you with meat of one kind or another. However, unless you have a gun, you will get most of your game via snares and traps, and most of what you will catch in this way will be small animals and birds.

If you do have a gun, observe the following rules when hunting prey: walk as

CANADIAN SPECIAL FORCES TIPS

WHAT TO LOOK FOR WHEN HUNTING

Canada's elite troops are adept at locating animals for food in the frozen wastes of their country by being alert and aware of animal habits.

- Animals themselves: they will be unsure when they see you. Remain still, make slow movements, if you have a gun make the first shot count.
- Trails beaten down by heavy usage.
- Tracks, which can provide information on the type, size, age and sex of animals.
- Droppings: they are a good indicator of animal type and size.
- Feeding grounds and water holes: they are good sites for hunting in the early morning or evening. Trails leading to them can be set with snares or traps.
- Dens, holes and food stores are good sites for setting snares.

quietly as possible; move slowly, stop frequently and listen; be observant; hunt upwind or crosswind whenever possible; blend in with the terrain features if you can. Be prepared: game often startles the hunter and catches him off guard, resulting in an ill-aimed shot.

TRAPPING ANIMALS

Mammals can be a valuable food source for the survivor. They are divided into the following groups:

WILD CATS: range in size from domestic-sized cats to lions and tigers. Avoid the big cats unless you have a gun. The smaller ones can be caught with powerful spring snares.

WILD DOGS: can be caught with snares, though they have a nasty bite. Beware.

BEARS: best to avoid. These strong, big creatures can outrun a horse over short distances and can kill a man easily. Keep away from bear cubs (mother bear will be nearby), and remember that a wounded bear is extremely dangerous. Hunt only if you have a gun, and make sure you kill with the first shot to the brain. Bears can be killed with deadfall and spear traps, but make sure your trap will kill. Bears really are not worth it as an easy food source.

WEASELS, STOATS, MINK, MARTENS AND POLECATS: beware, they have sharp teeth. They can be caught by spring snares and deadfalls.

WOLVERINES: badger-shaped animals. Only tackle if you have a gun. Can be caught with spring snares.

BADGERS: have a fierce bite. Can be caught in spring snares and deadfalls.

CATTLE: large cattle, especially bulls, can be dangerous. Use only very powerful snares, spring traps and deadfalls.

DEER AND ANTELOPES: beware of their horns, they can gorge and stab and inflict serious wounds. The small types can be caught in snares and deadfalls, and the larger ones by spear traps and larger deadfalls.

WILD PIGS: their tusks can cause serious injury. Can be caught with strong spring snares, deadfalls and pig spear traps.

RODENTS: rabbits are easily caught with snares and spring snares.

REPTILES: can be a valuable food source but some can be extremely dangerous and are best left alone: crocodiles, alligators, gila monsters and beaded lizards. Do not eat toads - they have toxic skin secretions.

SNAKES: treat all snakes as dangerous and poisonous, even if they are not. Use a forked stick to pin down just behind the head. Club back of the head with another stick. Better still, cut off the head with a machete. *NEVER* pick up a

CAUTION

Rabbit meat lacks vitamins and fats essential to man. When humans eat rabbit the body uses its own minerals and vitamins to digest it. They need to be replaced. If they are not, and you keep on eating rabbit, you can literally eat yourself to death.

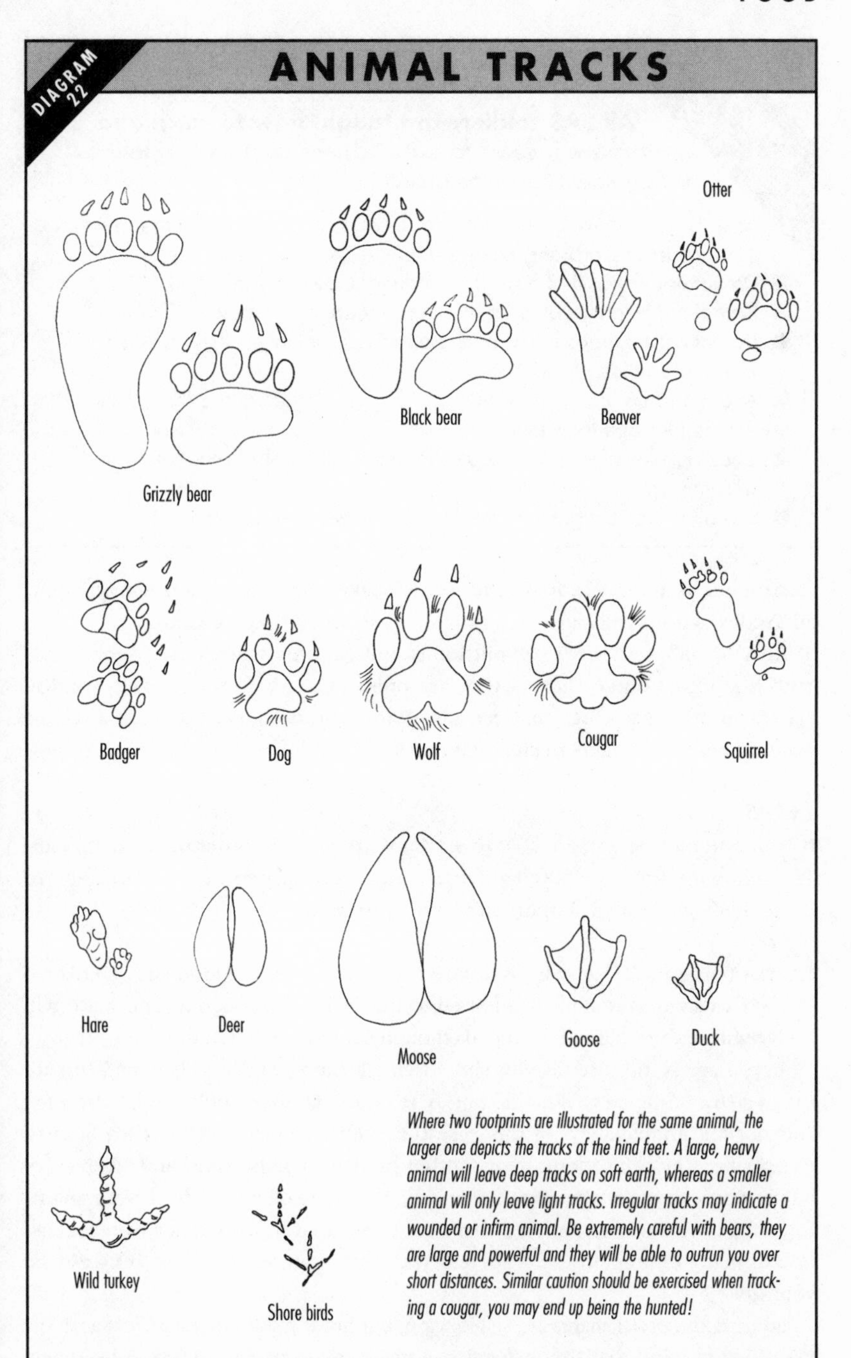

Where two footprints are illustrated for the same animal, the larger one depicts the tracks of the hind feet. A large, heavy animal will leave deep tracks on soft earth, whereas a smaller animal will only leave light tracks. Irregular tracks may indicate a wounded or infirm animal. Be extremely careful with bears, they are large and powerful and they will be able to outrun you over short distances. Similar caution should be exercised when tracking a cougar, you may end up being the hunted!

BRITISH SAS TIPS

EATING INSECTS

All SAS soldiers are taught how to catch and prepare insects to eat. Adhere to their guidelines when searching for insects.

- Be careful when searching for insects: their hiding places may also conceal scorpions, spiders and snakes.
- Do not eat insects that have fed on dung: they carry infection.
- Do not eat brightly coloured insects: they are poisonous.
- Do not collect grubs found on the underside of leaves: they secrete poisonous fluids.
- Avoid hornets' nests unless you are desperate: they guard their nests with vigour and their sting is vicious.
- Cook ants for at least six minutes to destroy the poisons that are found in some types.
- Boil all insects caught in water in case the water is polluted.

snake until you are sure it is dead, some snakes can feign death convincingly, and reflex actions can cause them to 'bite' you when they are dead.

SNAILS, WORMS AND SLUGS: nutritious, but eat only when fresh. Avoid snails with brightly-coloured shells - they are poisonous. Avoid sea snails in tropical waters, some have stings that can kill. With worms, starve them for a day or squeeze between fingers to clear out muck.

TRAPS

It is important that you are able to set traps correctly in order to catch animals. The following traps are all effective and relatively easy to make. Learn how to make them and, equally importantly, where to set them.

Snares (Diagram 23) A snare is a wire (A, D & E) or string loop (B & C) placed in such a way that an animal is forced to put its head through it. The snare will then tighten, thus killing the animal (though sometimes not immediately).

If you are setting snares you should check them regularly. It is unfortunate that a snare designed to strangle one type of animal may catch another by a leg and not kill immediately. In this case the animal may bite off its own limb to escape, or attract the attention of another predator and be killed and taken away before you can get to the trap. Either way, you should always check your snares for trapped animals and ensure that they are working. Remember, you are just one of many hunters in the wild, and you cannot afford to let any prey escape your grasp.

Some commercial snares are self-locking, but home-made snares are just as effective. Bear in mind that the material you make the snare out of has to be strong

Animal	Diameter loop	Height above trail
Hare	10cm (4.5in)	7.5cm (3in)
Squirrel	7.5cm (3in)	4cm (1.5in)
Rabbit	10cm (4in)	6cm (2.5in)
Fox	25cm (10in)	30cm (12in)
Wolf	40cm (16in)	45cm (18in)
Beaver	12cm (5in)	2.5cm (1in)

enough to catch its intended prey. The size of the loop needed to catch certain animals and the height of the snare is shown below:

Positioning snares (Diagram 24) Your aims when positioning snares are: keep the loop open and unrestricted so it can tighten on the animal, and keep it a proper distance off the ground. Remember, wire snares are easy to position because of their rigidity. Snares should be sited as follows:

☐ On heavily used trails (A, B & C) or in an area where animals are feeding on vegetation or on a carcass.

☐ Near a den or well-used stores of food (D).

Snaring a main trail is called a trail set and is an effective way of catching animals

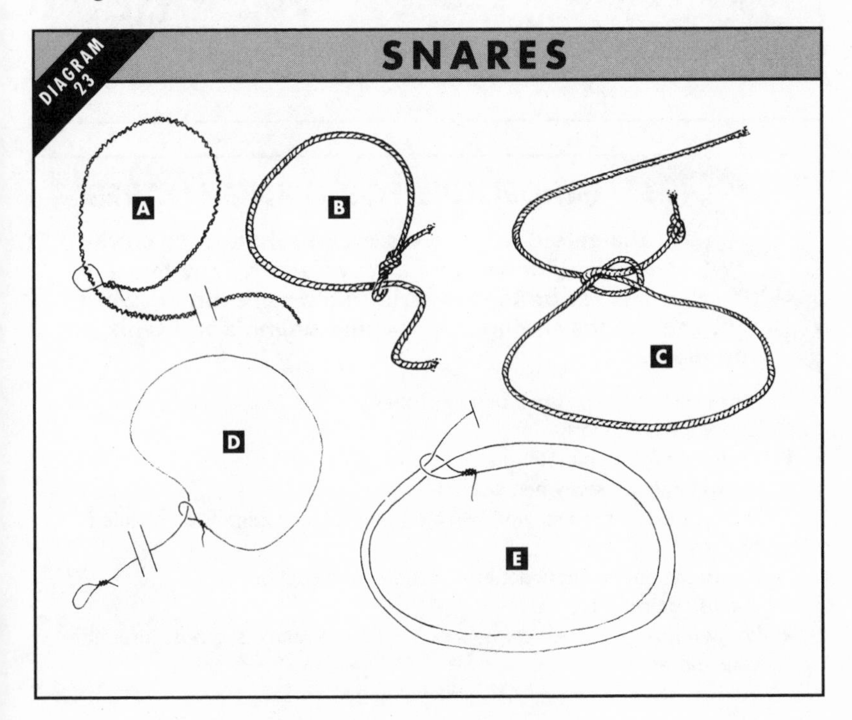

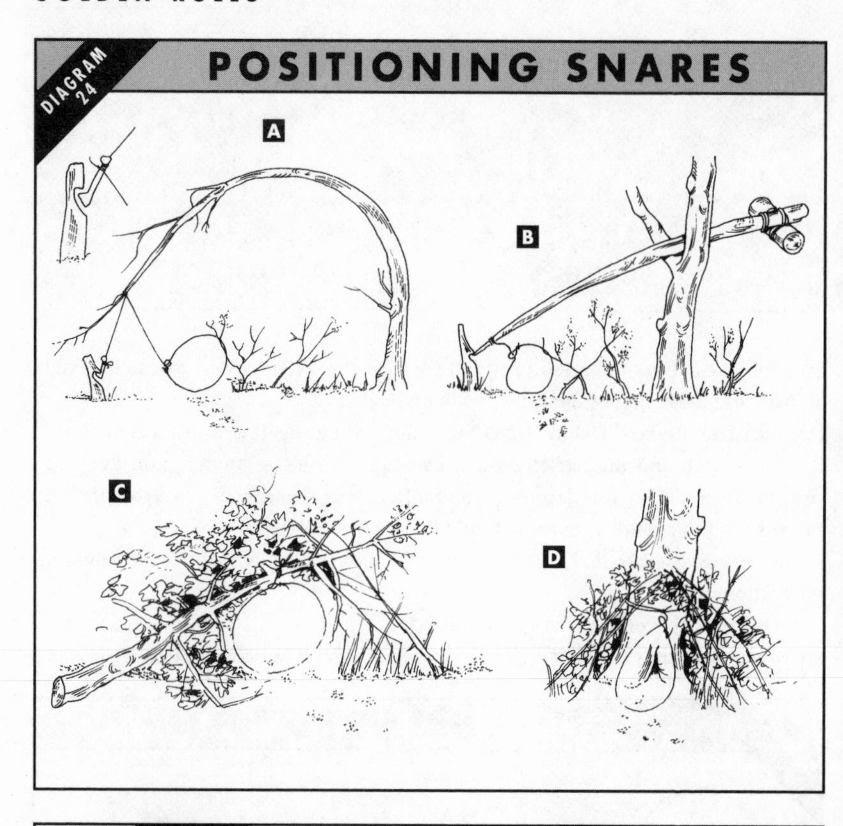

US SPECIAL FORCES TIPS

GUIDELINES FOR USING SNARES

The elite Green Berets are taught how to catch animals with the minimum of effort. They learn that it is far better to work with nature than against it. This means setting traps where animals will walk into them.

- Make sure the traps are working properly.
- Check them regularly.
- Do not walk on animal trails.
- Always lay trail sets when possible.
- Try to place a snare so that, when an animal is caught, it will be lifted off the ground.
- Approach any animal caught in a snare with caution.
- Use fish entrails as bait.
- Position foliage in such a way that it will force animals to pass through your snares.

(remember, animals are creatures of habit, they will stick to well-worn trails and so should you).

Deadfall traps (Diagram 25) The principle of these traps is simple: when the bait is taken a weight falls on the prey and kills it. There are many types of deadfall triggers, but they are all activated either by a tripline release action or a baited release action.

With the tripline release the animal touches or trips a line, stick or pole, activating the deadfall. With the baited release the animal is attracted to the deadfall by the bait, and when it pulls on it the deadfall will drop.

TAKE CARE

Deadfalls, especially the larger ones, can kill humans, and they can be set off easily. Remember where you set your traps.

Spear traps (Diagram 26) These traps can be very effective. They consist of a springy shaft that is held in place by a tripline, with a spear being firmly lashed to the springy shaft, which hits the animal when released. *WARNING: THESE TRAPS CAN KILL. ALWAYS APPROACH THEM FROM BEHIND.*

US ARMY TIPS

MAKING A FIGURE FOUR TRIGGER

The figure four trigger (see Diagram 25) is one of the easiest triggers to make. It is used in deadfall traps. Here's how to make it:

For the upright stick

- Cut the top at an angle and square off the tip to allow it to fit into notch in the release stick.
- Cut a square notch near the bottom to fit a corresponding square notch in the bait stick. Flatten the sides of the stick at this notch to guarantee a good fit with the bait stick.

For the release stick

- Cut the top so the deadfall will rest on it securely until triggered.
- Cut a notch near the top in which to fit the upright stick.
- Cut the bottom end at an angle to fit into the bait stick.

For the bait stick

- Cut a notch near one end in which to place the end of the release stick.
- Shape other end to hold the bait.
- Cut a square notch at the spot where it crosses the upright stick; the notches in the upright stick and the bait stick should fit firmly together.
- Rest the trigger on a stone or a piece of wood to stop it sinking into the ground.

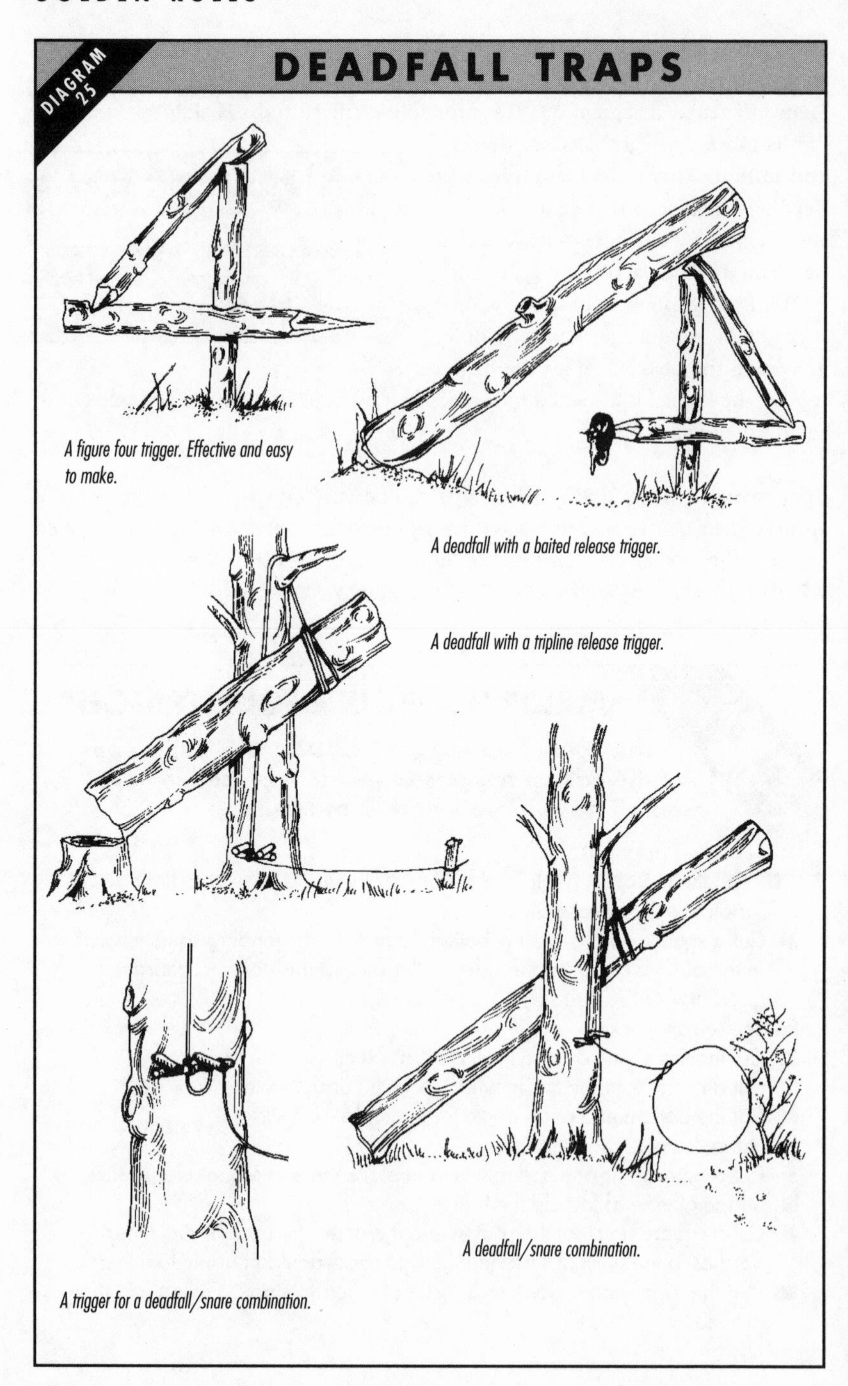

A figure four trigger. Effective and easy to make.

A deadfall with a baited release trigger.

A deadfall with a tripline release trigger.

A deadfall/snare combination.

A trigger for a deadfall/snare combination.

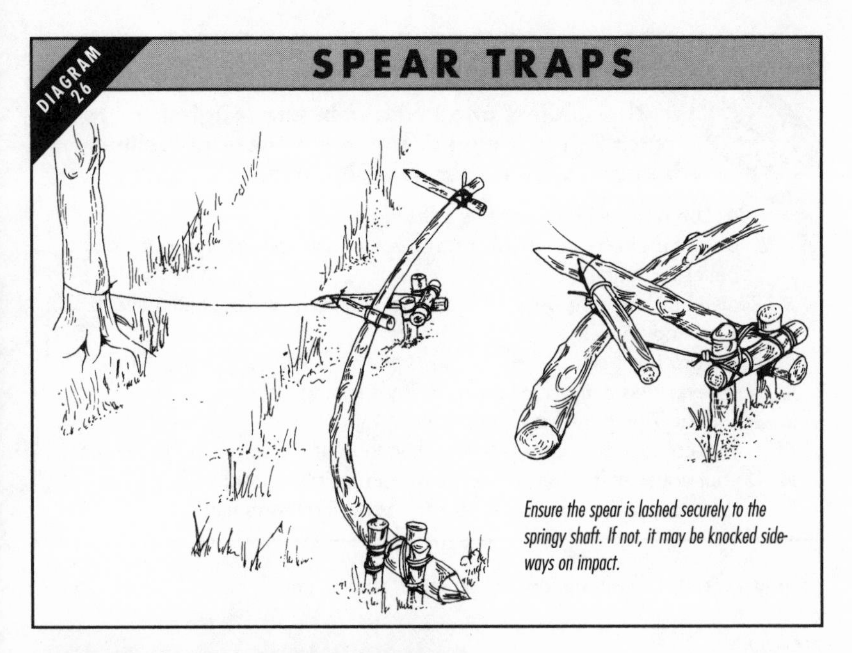

Ensure the spear is lashed securely to the springy shaft. If not, it may be knocked sideways on impact.

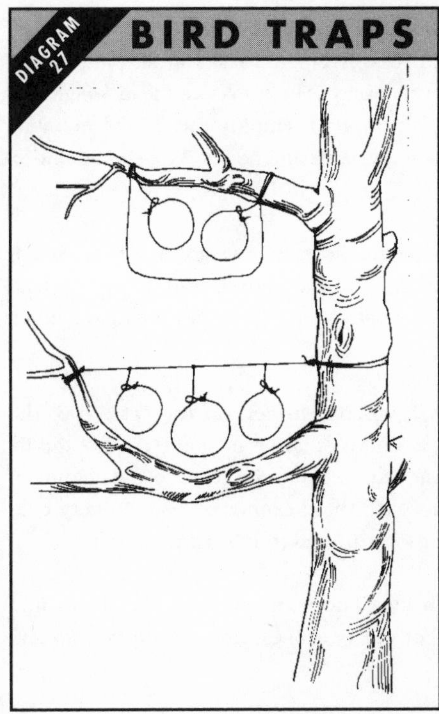

Bird traps There are several effective ways of catching birds. An extremely simple method is to put a stone in a piece of bait and throw it into the air. A bird will attempt to swallow it but the stone will catch in its mouth, causing it to fall to earth, whereupon you can club it. The following traps will also serve you well:

SUSPENDED SNARES: hang a line of snares across a stream above water level.

BAITED HOOKS: fish hooks buried in fruit or other food are a good way of catching birds. The hooks catch in the birds' throats.

NOOSE STICKS: (Diagram 27) tie a number of nooses 1.25-2.5cm (0.5-1in) in diameter close together along a branch or stick. Position in a favourite roosting or

US SPECIAL FORCES TIPS

FISHING TIPS

The soldiers of all elite units are taught how to catch fish in the wild. Try to use these guidelines when you are in a survival situation.

- Use natural bait whenever possible.
- Do not use too big a hook for the type of fish you are trying to catch, better too small than too big.
- Remember how you were fishing when you get a strike; continue with this method.
- If you are not catching fish, change your methods and/or your lures.
- Save eyes and entrails for the next day's fishing.
- Try spear fishing – it can be very effective.
- Fish become more active feeders when there is a change in the weather.
- Do not eat shellfish that are not covered by water at high tide.
- Mussels are poisonous in tropical zones during the summer.

nesting spot. Birds become entangled when they alight.

FISHING

Fishing (Diagram 28) can be an invaluable aid to the survivor when it comes to acquiring food. There are fish in the seas, rivers and lakes in all parts of the world, and they can be caught relatively easily. The hooks and lead weights in your survival kit are worth their weight in gold. Employ the fishing methods listed below to catch fish using your line, weights and hooks. You should improvise a gaff hook (H) to land large fish.

Still fishing (A & B) Weight your line with a float, lead weight or a rock. Attach baited hook or hooks and allow it to settle on the bottom of the river or float. Take up the slack and wait for a strike. Remember to pull on the line once in a while.

Dry fly fishing This method is used when fish feed off the surface of the water. Improvise a line with a stick, a length of string or line (G). Cast the fly upstream and let it float down past you. Experiment with the size and colours of your flies. Remember that this method of fishing cannot be used in very cold weather when there are no airborne insects around (the fish will not bite).

Set lining (D & E) Involves casting a long line with several baited hooks into the water and leaving it overnight. Put out two lines, one on the bottom and one off the bottom.

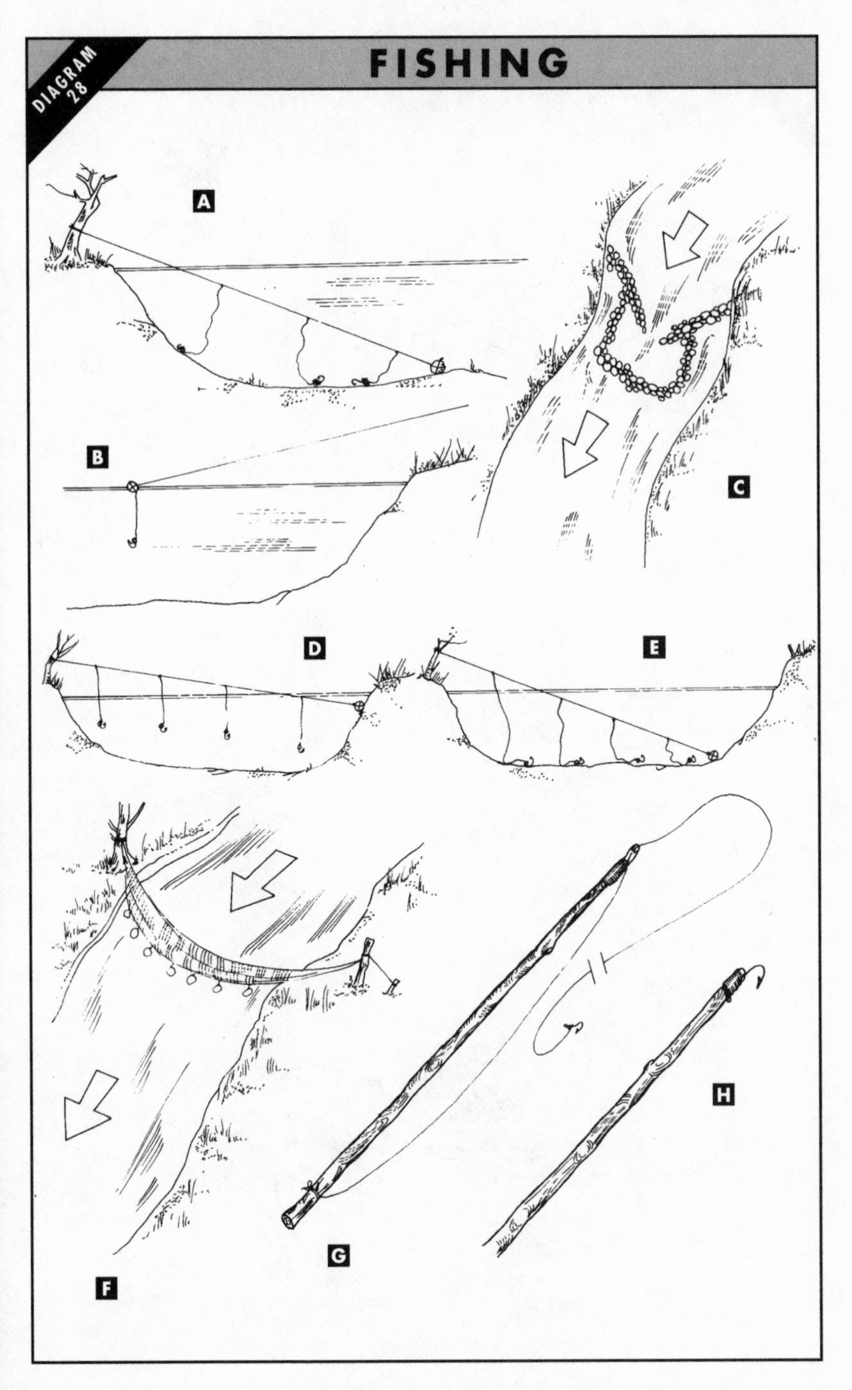
DIAGRAM 28
FISHING
A
B
C
D
E
F
G
H

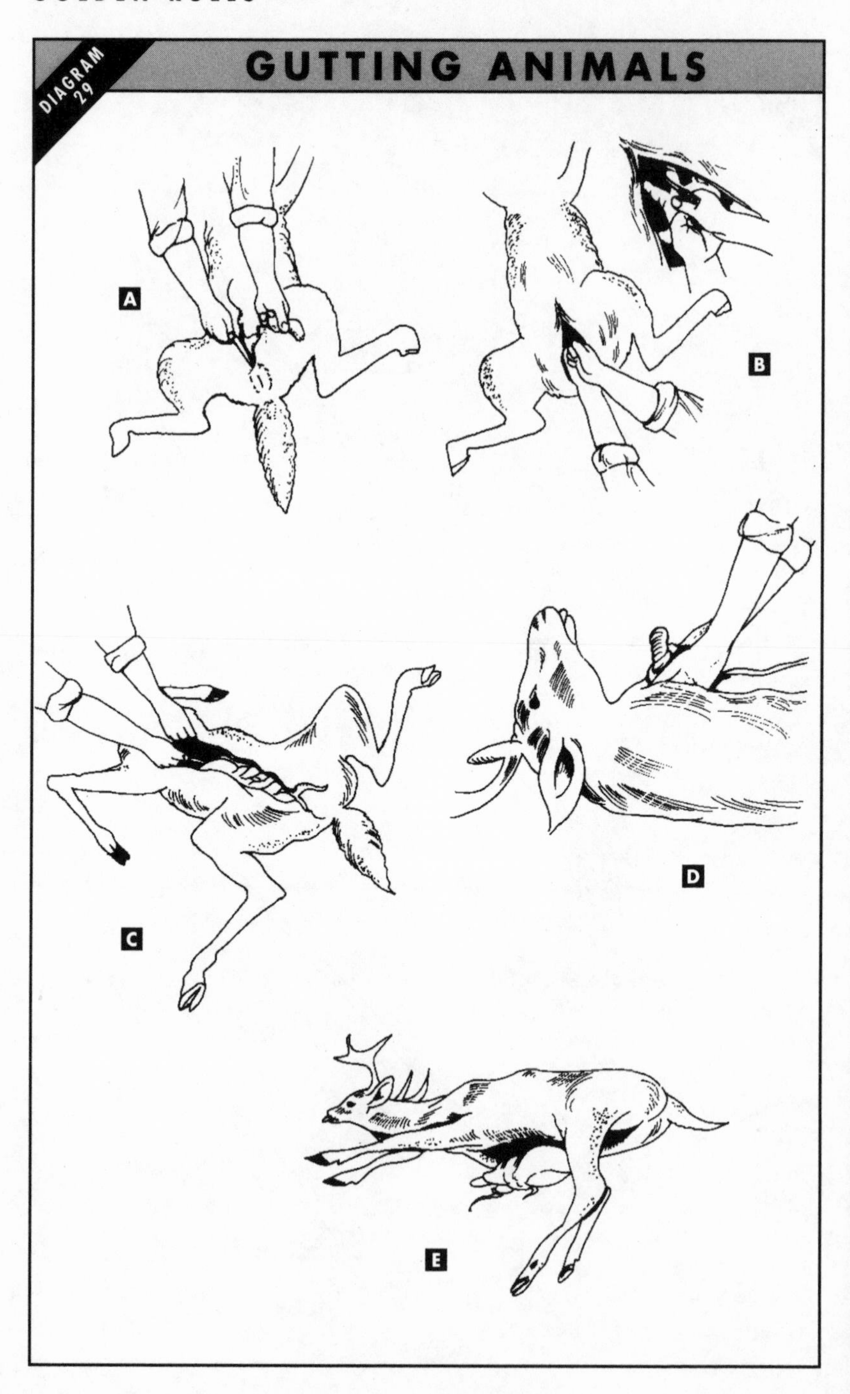
DIAGRAM 29
GUTTING ANIMALS
A
B
C
D
E

Gill netting (F) This is done by setting out a net that is constructed to catch fish by their gills as they try to swim through it. It is very effective for streams; tie stones along the bottom edge of the net to keep it on the bottom. Alternatively, make a trap with stones or rocks and herd fish into it (C).

GUTTING ANIMALS

After you have killed the animal, slit its throat to bleed it. Save the blood if you can, it is full of vitamins and minerals. Place the carcass, belly up, on a slope if possible (Diagram 29).

Gut around the anus, and if the animal is a male cut the skin parallel to, but not touching, the penis (A). Insert the first two fingers between the skin and the membrane enclosing the entrails. Place knife blade between the two fingers and extend the cut to the chin (B). Cut the diaphragm at the rib cage; cut the pelvic bone and remove the anus (C); and split open the breast and remove as much of the windpipe as possible (D). Turn the animal on its side and roll out the entrails (E).

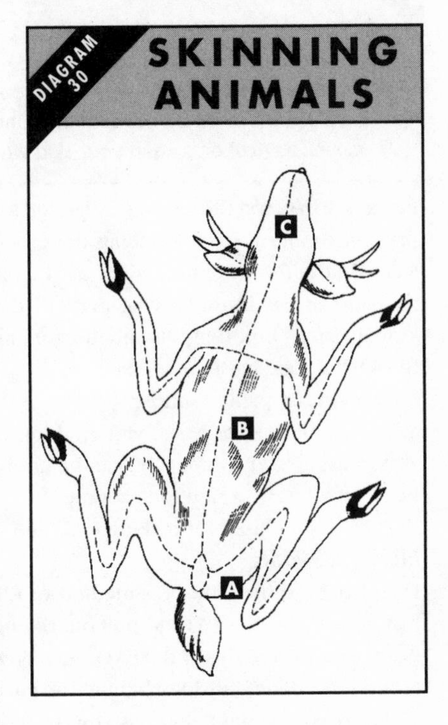

SKINNING

When you have caught the animal lay it on its back (Diagram 30) Cut the skin on a straight line (it will help if you have a sharp knife) from the tail bone to a point under the animal's neck (A-B-C). Press skin open until the first two fingers can be inserted between the skin and the thin membrane enclosing the entrails. Place knife blade between the fingers, blade up, and force fingers forward, palm upward, cutting the skin but not the membrane. See Diagram 29 for how to cut.

On reaching the ribs, you do not have to be as careful. Take away you fingers, force the knife under the skin and lift. Cut to point C. With the central cut completed, make side cuts running from the A-C line. Cut up the inside of each leg to the knee and hock joints. Then make cuts around the front legs just above the knees and around the hind legs above the hocks. Make the final cross at point C, and then cut completely around the neck and at the back of the ears. Now begin skinning.

Begin at the corners where the cuts meet. After skinning the animal's side as

BRITISH SAS TIPS

RULES FOR AVOIDING DISEASED ANIMALS

It is important to know if a beast is diseased. Follow British SAS rules concerning the identification of ill animals.

- Check lymph glands (in the cheek). If they are large and discoloured the animal is ill.
- Animals that are distorted or discoloured about the head are diseased; their meat should be boiled thoroughly.
- Cover any cut or sore in your skin when preparing diseased animals.

far as possible, roll the carcass on its side to skin the back. Spread out loose skin to prevent the meat from touching the ground and turn the animal on its skinned side. Follow the same procedure on the opposite side until all the skin is free.

You can use the following parts of the animal:- ENTRAILS, heart, liver and kidney can all be eaten, though discard any that are discoloured.

BLOOD: a good base for soups.

FAT: good for making soups.

SKIN: can be used as leather for clothing.

TENDONS AND LIGAMENTS: can be used for lashings.

BONE MARROW: a rich food source.

BONES: can be used for making tools and weapons.

FILLETING FISH

Diagram 31 shows the best method of filleting fish. Slit fish from the anus to just behind the gills (A) and pull out the internal organs (B). Wash and clean the flesh, trim off the fins and tail (C). Cut down to, but not through, the spine. Cut around the spine, finishing behind the gills on both sides (D).

Insert your thumb along the top of the spine and begin to pull it away from the flesh (E). Ribs should come out cleanly with the spine (F).

US ARMY RANGERS TIPS

GUIDE FOR PREPARING SNAKES FOR EATING

US Army Rangers are taught how to overcome their natural fear of snakes and turn the animal into a tasty and nutritious meal.

- Grip dead snake firmly behind the head.
- Cut off the head with a knife.
- Slit open belly and remove innards.
- Skin snake (use skin for improvising belts and straps).

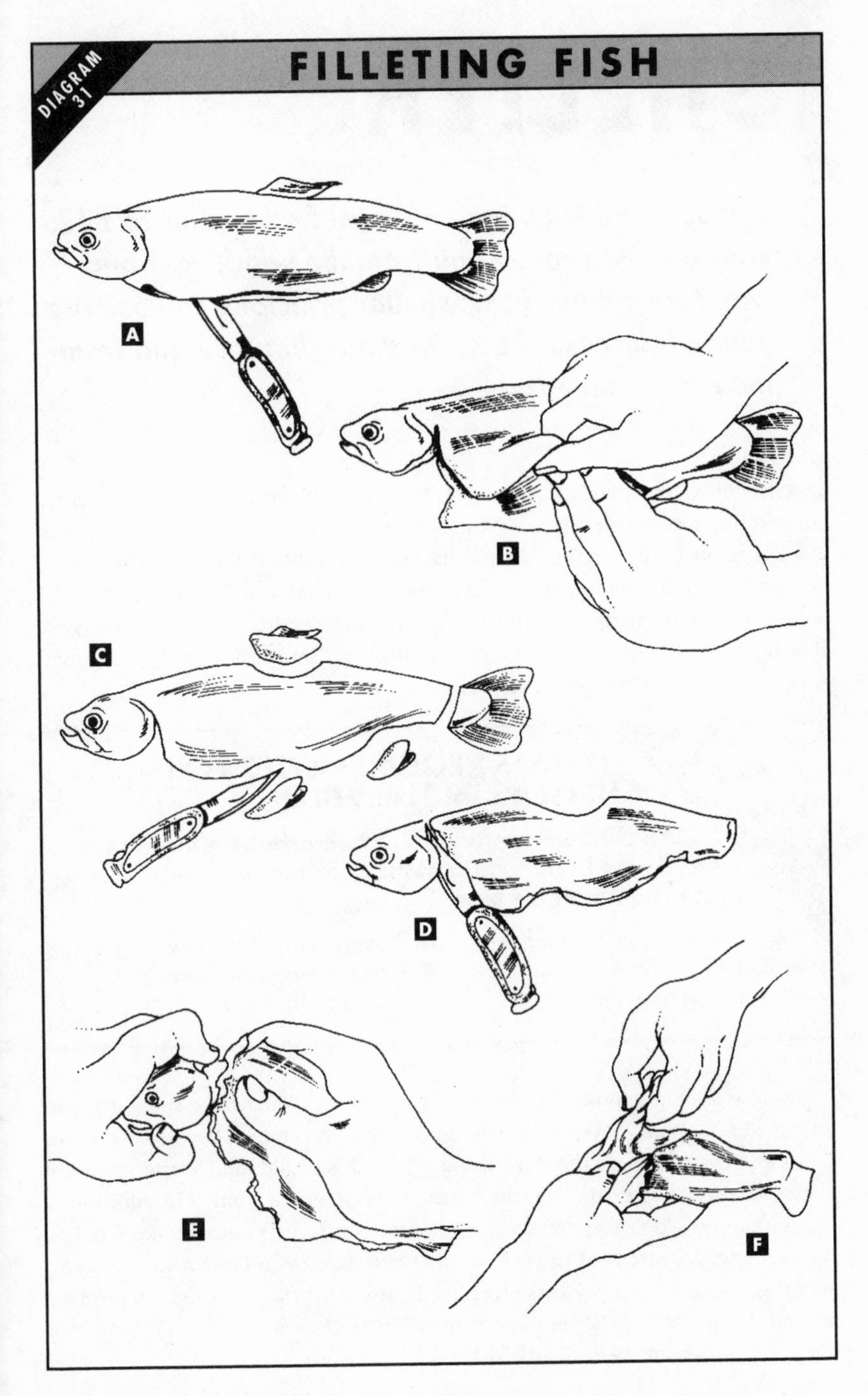
DIAGRAM 31
FILLETING FISH
A
B
C
D
E
F

SHELTER

In a survival situation you must find shelter or build your own to protect you from the wind, cold and wet. Learn these basic shelter principles to construct your wilderness shelter in the right place and from the proper materials.

Specific types of survival shelters will be dealt with in detail in the chapters concerning survival in different kinds of terrain. However, there are several general points with regard to shelter that you should bear in mind when you are in a survival situation, regardless of the environment you find yourself in. Of course, if you are suddenly trapped by the weather, or are injured or exhausted, almost any natural shelter will do. For example, get into a hollow in the ground and add to its height by piling up rocks around it.

BRITISH SAS TIPS

LIST OF REQUIREMENTS FOR SHELTERS IN THE WILD

SAS soldiers are taught to build shelters that are sturdy and will fulfil a number of requirements. These requirements are:

- Protection from the cold.
- Protection from the wind.
- Protection from insects.
- Protection from the snow.
- Protection from the damp.
- Protection from too much sun.

There is a temptation if you are in a warm and dry climate to assume that you will need less shelter, even no shelter at all. However, remember that whatever the temperature during the day, at night it will get cold, and warm areas are subject to changes in the weather just as much as cold ones. In addition, a shelter can provide protection against unforeseen threats and wildlife (snakes, for example, are attracted to body warmth and have been known to crawl into sleeping bags of backpackers sleeping in the open and curling themselves around the genitals of the slumbering occupant!). Therefore, do not believe that you can do without shelter in the wild.

BRITISH SAS TIPS

WHERE NOT TO BUILD A SHELTER

SAS soldiers often have to build shelters quickly when on operations behind enemy lines. They know they must avoid the following spots:

- On a hilltop exposed to wind: it will be cold and windy.
- In a valley bottom or deep hollow, they could be damp and are prone to frost at night.
- On a hillside terrace that holds moisture: they are invariably damp.
- On spurs of land that lead to water: they are often routes to animals' watering places.
- Below a tree that contains a bees' or hornets' nest or dead wood. Dead wood could come crashing down on you in the next high wind.
- Under a solitary tree: it can attract lightning.

CHOOSING A SITE

Choosing the right site for a shelter is very important. If you select a bad site you will probably end up building another shelter in a better spot, and will thus waste valuable time and energy. *DO NOT* select a site in the late afternoon after a long day's walk or march, ie at the last minute. You will be tired and in no mood to make a calculated assessment. Your decision will invariably be a bad choice and may force you into using poor materials.

WEATHER CONSIDERATIONS

The weather can play a key part in determining the location and type of shelter you build. For example, low areas in cold regions have low night temperatures and suffer from windchill. Valley floors invariably have colder temperatures than higher up - so-called 'cold air sumps'. Therefore, in cooler areas, try to situate your shelter where it can take advantage of the sun (if it comes out!), and remember to use plenty of insulation material.

In the desert your shelter must protect you from extremes of both heat and the cold, though damp will not be a problem.

Wind In warm areas, locate shelters to take advantage of breezes, but beware of exposing a shelter to blowing sand or dust, both of which can cause injury and damage. In cold regions, choose a site that is protected from the effects of wind-chill and drifting snow.

Rain, sleet and snow can all be potential hazards. Do not build a shelter in a major drainage route, in a site that is prone to flash floods or mud slides, nor in an avalanche area.

INSECTS

Insects can be a problem around a camp. If you build your shelter where there is a breeze or steady wind you can reduce the number of insects that will pester you. Avoid building a shelter near standing water, as it attracts mosquitoes, bees, wasps, and hornets, and do not erect a shelter on or near an ant hill, unless you want a never-ending succession of bites and stings.

When you are building your shelter look above you. The tree you are under may contain a bees' or hornets' nest, which you will obviously want to avoid. In addition, watch out for dead wood in trees above you. It make be firmly in place at the moment, but in the next storm or high wind it could come crashing down on you and your shelter.

IMPORTANT

Keep your shelter dry by constructing a run-off drainage channel gouged from the earth around the shelter. Ensure the shelter itself has plenty of ventilation, especially if you intend to light a fire for cooking inside it.

TYPE OF SHELTER

This will be determined by the conditions and range of materials available. Remember, it is often a good idea to improvise a makeshift shelter to give you some form of protection from the elements until you can erect something more permanent (particularly if it is getting dark and cold).

If you can't find any materials, make use of natural shelters, such as cliff overhangs and gradients. In open areas, sit with your back to the wind and pile your equipment behind you as a windbreak. The following are all examples of natural shelters that the survivor can use in an emergency situation:

- ☐ Brambles and boughs that sweep down to the ground or are partly broken. Add branches to make them more dense.
- ☐ Natural hollows provide protection from the wind, though be sure to divert downhill flow of water around them. Use a few strong branches, covered with sticks and turf, as a roof.
- ☐ Fallen tree trunks. Scoop out a hollow on the leeward side and cover with boughs to make a roof.
- ☐ Stones or small rocks can be used to increase the height of your hollow. To insulate against the wind, plug the gaps between the stones with turf and foliage mixed with mud.
- ☐ Caves can make excellent shelters. If it is a cave in a cliff or mountain, increase your cave's warmth by building a windbreak over the entrance. You can use stones, rocks or turf cut like bricks. If you intend to light a fire in a cave, remember to light it at the back, a fire at the front of a cave will eddy about inside and will choke you!

There are, of course, an almost limitless variety of shelters you can build in the wild. Use your imagination, but do not forget the basic principles.

ROPES AND KNOTS

All survivors must have a knowledge of ropes and basic knots and where and how to use them. Do not neglect rope craft and knots – they will both serve you well in a survival situation, and may even save your life.

It is important that the survivor has a basic knowledge of ropes and knots. It will be a great help in many situations, such as when building shelters, assembling packs, providing safety devices, improvising tools and weapons and even for first aid. It is very important that you practise the knots below before you need them: do not wait until you need them.

ROPES

Traditional rope materials include hemp, coconut fibre, Manila hemp and sisal, though rope can be made from any pliable fibrous material that produces strands of sufficient length and strength.

Many modern ropes are made from nylon and other man-made materials. These ropes are strong, light and resistant to water, insects and rot. However, they do have some drawbacks: they can melt if subjected to heat, they are slippery when wet, and they can snap if subjected to tension over a cliff edge. You should bear all these factors in mind when choosing rope.

ROPE TERMINOLOGY

Get to know the following words and phrases, they will help you greatly when tying knots:

BEND: used to join two ropes together or to fasten a rope to a ring or loop.

BIGHT: a bend or U-shaped curve in a rope.

HITCH: used to tie a rope around a timber or post so it will hold.

KNOT: interlacement of ropes or line to form a tie or fastening.

LINE: a single thread, string or cord.

LOOP: a fold or doubling of a rope, through which another rope can be passed.

A temporary loop is made by a knot or a hitch. A permanent loop is made by a splice.
OVERHAND LOOP OR TURN: made when the running end of the rope passes over the standing part.
ROPE (also called a line): made of strands of fibre twisted or braided together.
ROUND TURN: same as a turn, with the running end leaving the circle in the same general direction as the standing part.
RUNNING END: the free, or working, end of the rope.
STANDING END: the balance of the rope, excluding the running end.
TURN: describes the placing of a rope around a specific object, with the running end continuing in the opposite direction to the standing part.
UNDERHAND TURN OR LOOP: made when the running end passes under the standing part.

US ARMY RANGERS TIPS

THE CARE OF ROPES

America's reconnaissance specialists adhere to the following guidelines when using ropes for their mountain operations.

- Do not step on rope or it along the ground.
- Keep away from sharp corners or edges of rock, which can both cut rope.
- Keep rope as dry as possible, and dry it out if it becomes wet to avoid rotting.
- Do not leave rope knotted or tightly stretched longer than necessary, and do not hang it on nails.
- Exercise care with nylon rope: the heat generated by rope friction can often melt the fibres.
- Inspect rope regularly for frayed or cut spots, mildew and rot. If such spots are found, the rope should be whipped (prevented from unravelling) on both sides of the bad spots and then cut.
- Climbing ropes should never be spliced.

KNOTS

It is important to select the right knot for the right task. The knots listed below will be invaluable to you in a survival situation. Practise tying them, and remember to learn to untie them too (do not be caught in a situation where you need to untie a knot quickly but cannot).

There are four basic requirements for knots: they must be easy to tie and untie; they should be easy to tie in the middle of a length of rope; they can be tied when the rope is under tension, and they can be tied so that the rope will not cut itself when under strain. The following knots fulfil these criteria.

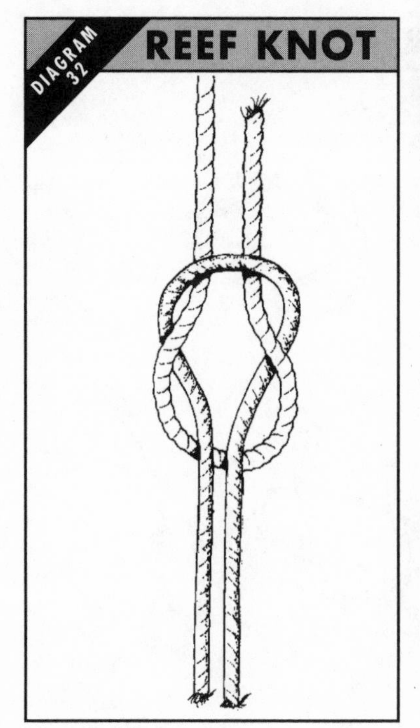

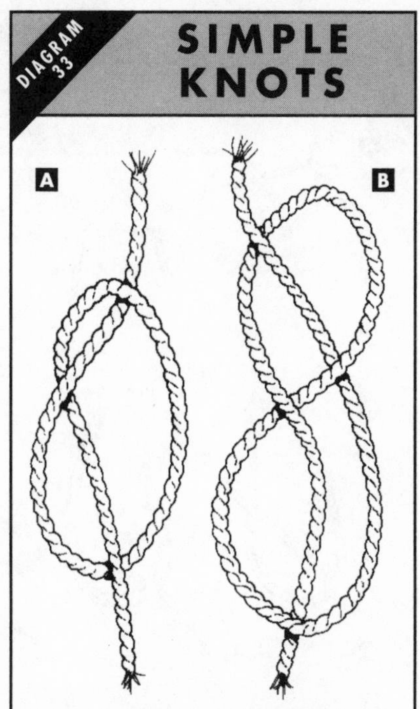

Reef knot (Diagram 32) This is the same as the square knot (see below), but note that it can also be tied by making a bight in the end of one rope and feeding the running end of the other rope through and around this bight. The running end of the second rope is threaded from the standing part of the bight. If the procedure is reversed, the resulting knot will have a running end parallel to each standing part, but the two running ends will not be parallel to each other. This knot is called a 'thief knot'.

Overhand knot (Diagram 33A) Of little use on its own, except to make an end-stop on a rope and to prevent the end of a rope from untwisting, but it does form a part of many other knots. Tie by making a loop near the end of the rope and passing the running end through the loop.

Figure-of-eight (Diagram 33B) Used to form a larger knot than would be formed by an overhand knot at the end of a rope. It is used at the end of a rope to prevent the ends from slipping through a fastening or loop in another rope.

To tie, make a loop in the standing part of the rope and pass the running end around the standing part back over one side of the loop and down through the loop. Then pull the running end tight.

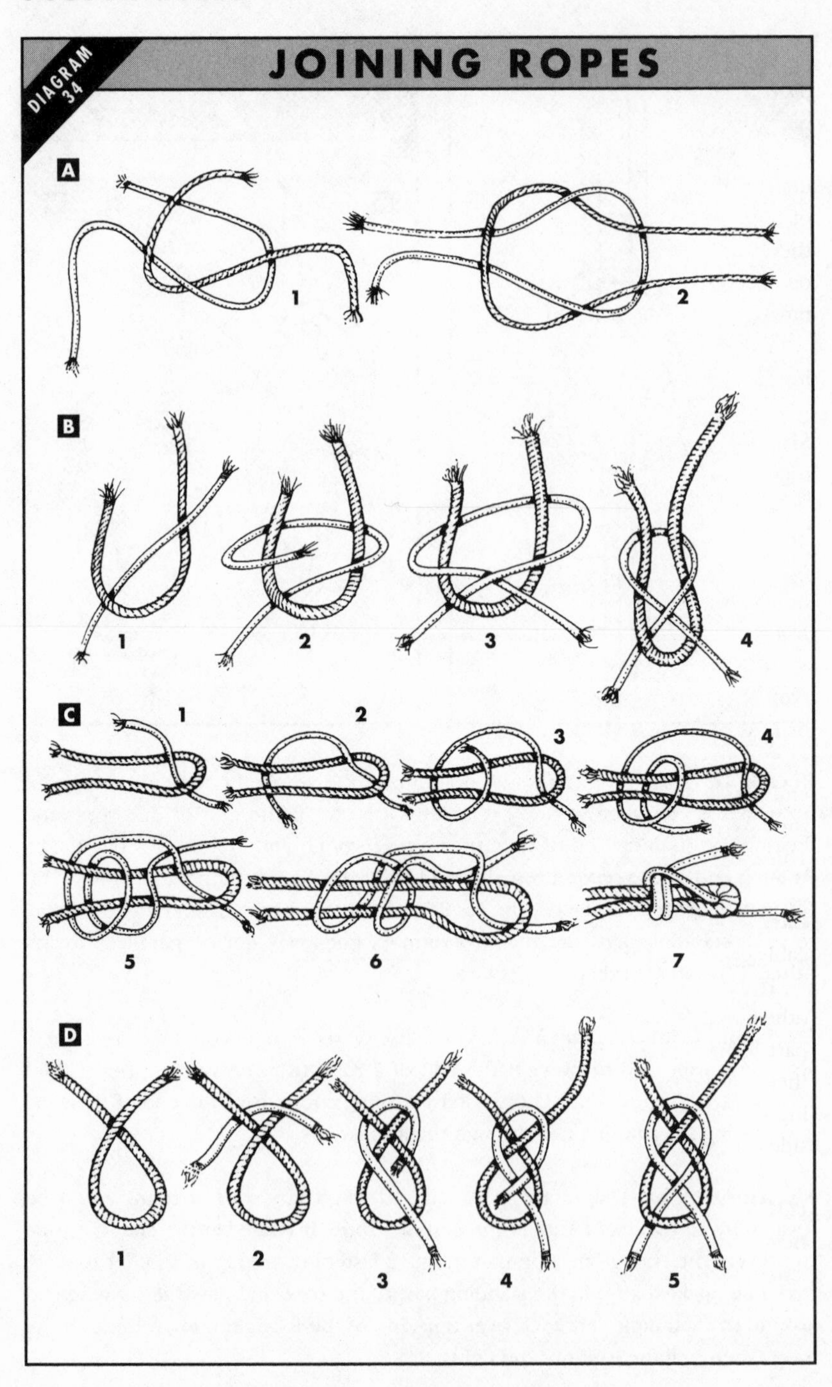
DIAGRAM 34
JOINING ROPES
A
1
2
B
1
2
3
4
C
1
2
3
4
5
6
7
D
1
2
3
4
5

Square knot (Diagram 34A) Used for tying two ropes of the *SAME DIAMETER* together to prevent slippage. A square knot should not be used for ropes of different diameters nor for nylon rope (it will slip). This knot is good for first aid because it will lie flat against the patient.

To tie, lay the running ends of each knot together but pointing in opposite directions. The running end of one rope can be passed under the standing part of the other rope. Bring the two running ends up away from the point where they first crossed and cross again (1). Once each running end is parallel to its own standing part, the two ends can be pulled tight (2). But note that each end must come parallel to the standing part of its own rope.

A square knot will draw tighter under strain. It can be untied easily by grasping the bends of the two bights and pulling them apart.

Single sheet bend (Diagram 34B) Used for tying two ropes of *UNEQUAL SIZE* together.

To tie, pass the running end of the smaller rope through a bight in the larger one (1). The running end should continue around both parts of the larger rope (2) and back under the smaller rope (3). The running end can then be pulled tight (4). *THIS KNOT WILL DRAW TIGHT UNDER LIGHT LOADS BUT MAY LOOSEN OR SLIP WHEN THE TENSION IS RELEASED.*

Double sheet bend (Diagram 34C) Used for joining together ropes of *EQUAL* or *UNEQUAL SIZE*, wet ropes, or for tying a rope to an eye.

To tie, tie a single sheet bend first (1-5). However, do not pull the running end tight. One extra turn is taken around both sides of the bight in the larger rope with the running end of the smaller rope (6). Then tighten the knot (7). This knot will not slip or draw tight under heavy loads.

Carrick bend (Diagram 35D) Used for heavy loads and for joining together thin cable or heavy rope. It will not draw tight under a heavy load.

To tie the Carrick bend, form a loop in one rope (1). The running end of the other rope is passed behind the standing part (2) and in front of the running part of the rope in which the loop has been formed. The running end should then be woven under one side of the loop (3), through the loop over the standing part of its own rope (4), down through the loop and under the remaining side of the loop (5).

LOOP MAKING

Bowline (Diagram 35A) Used to form a loop in the end of a rope. This loop is extremely easy to untie.

To tie, pass the running end of the rope through the object to be fixed to the bowline and form a loop in the standing part of the rope (1). The running end is

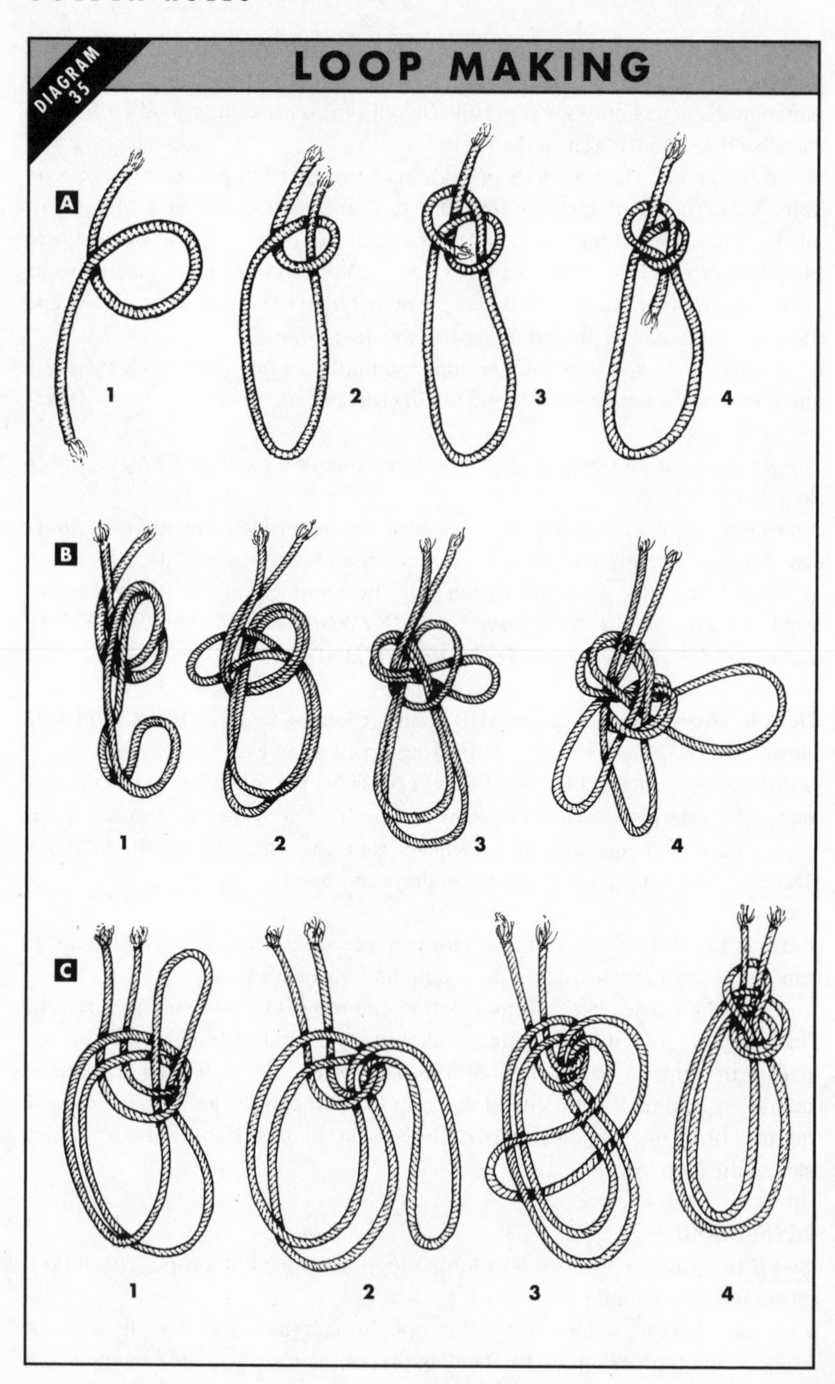
DIAGRAM 35
LOOP MAKING
A
1
2
3
4
B
1
2
3
4
C
1
2
3
4

then passed through the loop from underneath (2) and around the standing part of the rope (3) and back through the loop from the top (4). The running end passes down through the loop parallel to the rope coming up through the loop. Then pull the knot tight.

Triple bowline (Diagram 35B) Used as a sling or boatswain's chair (add a small board with notches for a seat). In addition, it can be used as a chest harness or as a full harness.

To tie, bend the running end of a line back to approximately 3m (10ft) along the standing part (1). The bight is formed as the new running end, and a bowline tied as described for the bowline knot (2-4). As a sling, the new running end, or loop (on the right), is used to support the back and the remaining two loops support the legs.

Bowline-on-a-bight (Diagram 35C) Used to form a loop at a point along a rope's length, rather than at the end. Easily tied and a knot that will not slip, it can also be tied at the end of the rope by doubling the rope for a short section.

When tying, a doubled portion of the rope is used to form a loop as for a bowline (1). The bight end of the doubled portion is passed through the loop, back down (2), up around the entire length of the knot (3) and tightened (4).

HITCHES

Hitches (Diagram 36) are used for attaching ropes to poles, posts and bars. Some of the most widely used and useful are give below. Try to master these hitches, they will serve you well in a survival situation.

Half hitch (Diagram 36A) Used to tie a rope to a piece of timber or another larger rope. *IT IS NOT A VERY SECURE KNOT OR HITCH.*

To tie, pass the rope around the timber, bringing the running end around the standing part and back under itself.

Timber hitch (Diagram 36B) Used for moving heavy timber or poles.

To tie, turn the running end about itself at least another time. These turns must be taken around the running end itself or the knot will not tighten against the pull.

Timber hitch and half hitch (Diagram 36C) Used to get a tighter hold on heavy poles for lifting or dragging.

To tie, pass the running end around the timber and back under the standing part to form a half hitch (see above). Tie a timber hitch further along the timber with the running end. The strain is on the half hitch, the timber hitch prevents the half hitch from slipping.

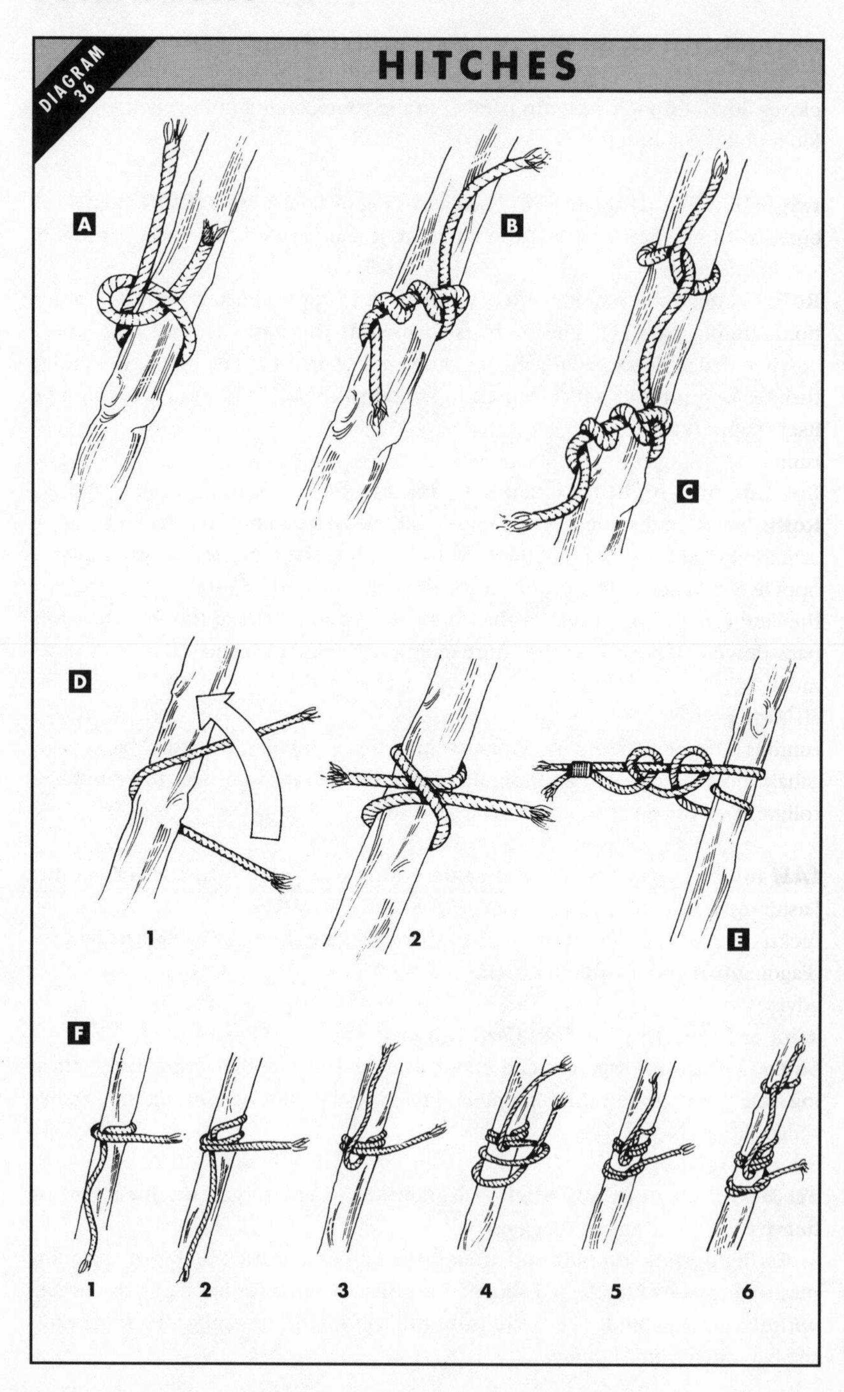
DIAGRAM 36
HITCHES
A
B
C
D
1
2
E
F
1
2
3
4
5
6

Clove hitch (Diagram 36D) Used to fasten a rope to a timber, pipe or post.

To tie in the centre of the rope, make two turns in the centre of the rope close together (1). Twist them so the two loops lie back-to-back. These two loops are slipped over the timber or pipe to form the knot (2).

To tie a clove hitch at the end of the rope, pass the rope around the timber in two turns so that the first turn crosses the standing part and the running end comes up under itself on the second turn.

Round turn and two half hitches (Diagram 36E) Used to fasten a rope to a pole, timber or spar.

To tie, pass the running end of the rope around the pole in two complete turns. The running end is brought around the standing part and back under itself to make a half hitch. Make a second half hitch. For increased security, the running end of the rope should be secured to the standing part.

Rolling hitch (Diagram 36F) Used to secure a rope to a pole so it will not slip.

To tie, the standing part of the rope is placed along the pole in the direction opposite to the direction the pole will be moved (1). Two turns are taken with the running end around the standing part and the pole (2). Reverse the standing part of the rope so that it is leading off in the direction in which the pole will be moved (3).

Take two turns with the running end (4). On the second turn around, the running end is passed under the first turn to secure it (5). To make it secure, tie a half hitch with the standing part of the rope at least 33cm (1ft) along the rolling hitch (6).

LASHINGS

Lashings (Diagram 37) are useful in the construction of shelters, equipment racks, rafts and other structures. The most commonly used are the square lash, diagonal lash and shear lash. They are fairly simple to tie, but you are strongly advised to practise making them.

Square lash (Diagram 37A) Used to secure one log at right angles to another log. It is useful for building shelters.

To tie, tie a clove hitch around the log immediately under the place where the crosspiece will be located (1). In laying the turns, the rope goes on the outside of the previous turn around the log (2). Keep the rope tight. Three or four turns are necessary.

Carry the rope over and under both logs in an anti-clockwise direction. Make three or four circuits and then make a full turn around a log and a circuit in the opposite direction (3). Follow with a clove hitch around the same log that the lashing was started on (4).

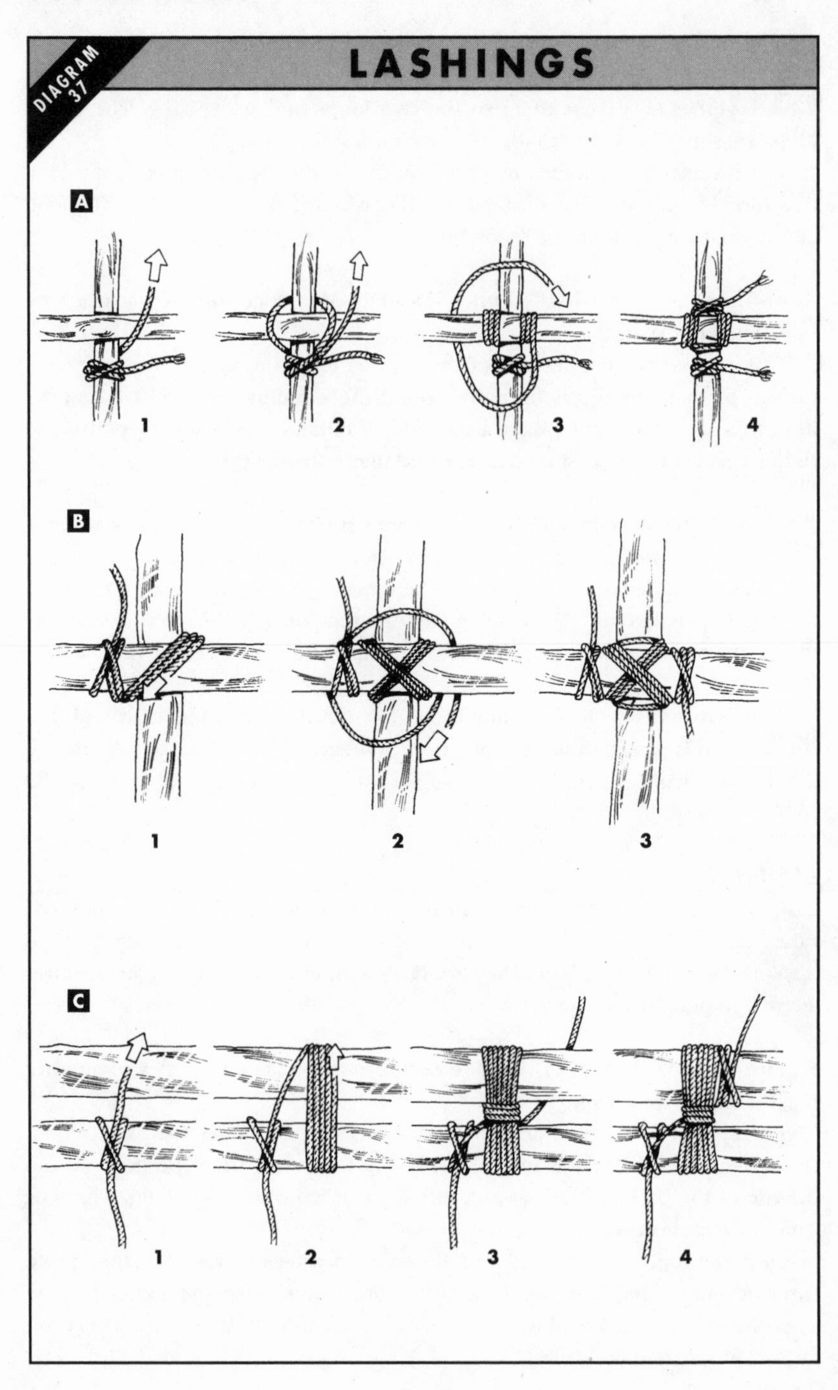
DIAGRAM 37
LASHINGS
A
1
2
3
4
B
1
2
3
C
1
2
3
4

Diagonal lash (Diagram 37B) Used to secure one log at right angles to another. It is an alternative to square lashing, and is much more effective when the spars do not cross at right angles, or when the spars are under considerable strain and have to be pulled towards one another to be tied.

Tie a clove hitch around the two logs at the point of crossing. Three turns are taken around the two logs (1). The turns lie beside each other, not on top of each other. Three more turns are made around the two logs, this time crosswise over the previous turns. Pull the turns tight.

Two frapping (diagonal) turns are made between the two logs, around the lashing turns (2). The lashing is finished with a clove hitch around the same pole the lash was started on (3).

Shear lash (Diagram 37C) Used for lashing two or more logs together.

To tie, place desired number of logs side by side. Start the lash with a clove hitch on the outer log (1). Logs are then lashed together using seven or eight turns of the rope loosely laid beside each other (2). Make frapping turns between each log (3). The lashing is finished with a clove hitch on the log opposite where the lash was started (4).

You can use a shear lash to make a tripod. Get three poles and make turns around all three and frapping in the two gaps. You can then position the bottoms of the poles to make a sturdy tripod. Alternatively, use a shear lash to make an A-frame using two poles secured with a shear lash, though make sure the feet are anchored in or on the ground to stop them spreading. It is very important to check that poles are secured correctly if they are under strain, and do not forget to check the rope or lashing material regularly for wear.

US ARMY TIPS

MAKING LASHINGS FROM TENDONS AND RAWHIDE

The US Army teaches its recruits to make lashing material from a dead animal. Tendons and rawhide make excellent lashing materials.

From tendons

- Remove tendons from game the same day it is caught.
- Smash dried tendons into fibres.
- Moisten fibres and twist them into a continuous strand; braid the strands if you require stronger lashing material.

From rawhide

- Skin the animal and remove all fat and meat from the skin.
- Spread out skin and remove all folds.
- Cut skin into strips.
- Soak the strips in water for 2-4 hours until soft and pliable.

IMPROVISING TOOLS AND WEAPONS

From the raw materials found in their surroundings, imaginative survivors should be able to make tools, weapons, utensils and even clothing, which will often serve much better than any commercially-made items.

Being able to fashion clothing and tools from the materials around you is a great bonus in a survival situation. Most of what you will make will be simple tools and weapons, but they will make your task of staying alive much easier. In addition, being able to make what you need will be a great morale-booster. If a tool or weapon breaks it can be a tremendous blow to your morale, but your optimism will return quickly if you know you can replace it with an improvised article. In a survival situation the number of things you can improvise is limited only by your imagination.

CLOTHING

If you are involved in an air crash, in the short term you should try to salvage as much as you can from the crash site, e.g. towels, tablecloths, curtains, cushions, and seat covers. Almost any type of fabric can be used for bedding, garments or shelter - be imaginative.

Insulation You can increase insulation by wearing layers of clothing (see What to Wear Chapter). Wear one sock on

CATTAIL - THE WORLDWIDE INSULATOR

Cattail grows everywhere in the world with the exception of the forested regions of the far north. It is a marshland plant found along lakes, ponds and backwaters of rivers. The fuzz on the top of the stalks forms dead air and makes a down-like insulation when placed between two pieces of material.

BRITISH SAS TIPS

CLOTHING TIPS

The uniforms of elite soldiers on extended operations can often wear thin. They are therefore taught how to improvise items of clothing.

- Tie long leaf strips and fibres around a belt or neck band to create a grass skirt or cape.
- Cut a head hole in a blanket or carpet to make a poncho. Tie at waist.
- It is easier to sew together or thong small pieces of hide. Fur on the inside gives better insulation.

top of another. Put dry grass or moss between them. Leaves from deciduous trees (those which lose their leaves each autumn) make good insulation. The leaves should be placed between two layers of material and stuffed in cuffs and waist bands. Paper, feather, animal hair and down can also be used for insulation.

Keeping out the wet Waterproofs can be fashioned from plastic bags or sheets. In addition, cut off large sections of birch bark; cut off the outer bark and insert the soft and pliable inner bark under the under-clothing - it will protect you from rain (you can also use other smooth barks that peel easily).

Rub animal fat into clothing, though not in conditions of extreme cold - the reduction in insulation is too great.

Footwear Shoe soles can be made from rubber tyres. Insert holes around the edges for thongs to tie them over wrapped feet or to sew onto fabric uppers (remember that several layers of wrapping are better on the feet than one).

Fashion a pair of moccasins from a piece of leather. Place your foot on the leather and cut out a piece 8cm (3in) bigger all round than the sole of your foot. Thong in and out around the edges and gather them over wrapped feet. Tie off the gathering thongs and weave another thong back and forth over the foot to make more secure.

TOOLS AND WEAPONS

Club Despite its simplicity, this is probably one of the most useful tools you can have in a survival situation. It is easy to make and can be replaced with minimum effort. Whatever environment you are in, try to make a club. It will be worth its weight in gold.

Fashion a club from a branch 5-6cm (2-2.5in) in diameter and around 75cm (2.5ft) long. It can be used for checking snares and deadfall traps, finishing off a trapped animal, and as a weapon for killing slow-moving game.

US ARMY TIPS

MAKING A WEIGHTED CLUB

One of the most hi-tech armies in the world teaches its recruits how to make a primitive tool. A weighted club can be used for breaking and hammering or for killing small animals.

- Do not select items that are too big or heavy.
- Find a stone that is shaped to allow you to lash it securely to the club, e.g. a stone with an hourglass shape.
- Find a piece of wood (straight-grained hardwood is best).
- Lash stone to the handle (you can split the handle, or shave the end of the club to half its diameter and then wrap the shaved end over the stone and lash securely).
- Inspect lashings regularly for wear and tear.

Wire saws (Diagram 38) Improvise a saw using a green sapling (A), or make a bucksaw (B) if you feel more ambitious. You can use the wire saw in your survival tin for the cutting edge (see What to Carry Chapter).

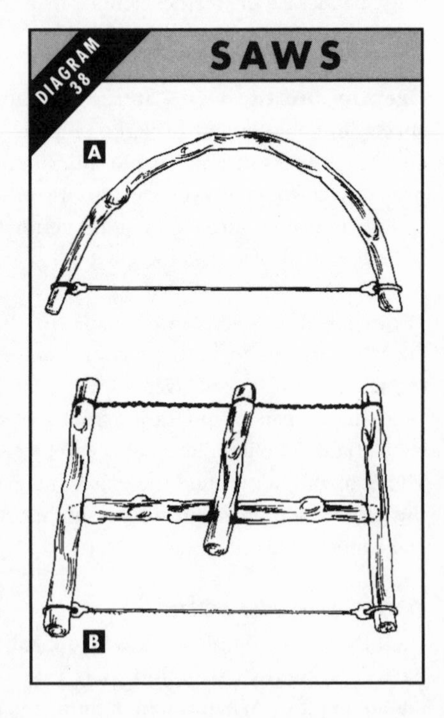

Stone tools Flint, obsidian, quartz, chert and other glassy tools can be used by the survivor. Stones can make good hammers either used on their own or lashed to a handle. Glassy stones can be chipped and flaked to make a sharp edge. When chipping with another stone the blow should be at an angle of less than 90 degrees, otherwise the shock will be absorbed within the stone. It takes practice to make stone tools.

Bone tools If you have killed a large animal do not discard the carcass. Antlers and horns can be used for digging, gouging and hammering. You can use a knife to carve bone. For example, a shoulder blade can be split in half and then teeth cut along it – then you have a saw. Even the bones from small animals can be useful: ribs can be sharpened into points; other bones can be sharpened and the other end burnt through with hot wire to produce an eye – the result is a sturdy needle.

Bow and arrows (Diagram 39) Tension in unseasoned wood is short-lived – make several bows and change when the one you are using loses its spring. Yew is best but you may have to use other wood depending on the area you are in. The stave should be about 120cm (4ft) long. Shape it so it is 5cm (2in) wide at the centre and tapers to 1.5cm (5/8in) at the ends. Notch the ends 1.25cm (0.5in) in to take the string. Rub bow all over with oil or animal fat.

For the string, rawhide is best, though any string or coil will do. When strung the string should only be under slight tension (A): you provide the rest when you pull to shoot. Secure string to the bow using a round turn and two half hitches (see Ropes and Knots Chapter).

Arrows should be made from straight wood 60cm (2ft) long and 6mm (0.25in) wide. They should be as smooth and straight as possible. Notch one end 6mm (0.25in) deep to fit the bow string. Arrow flights can be made from feathers, paper, light cloth or leaves trimmed to shape. If you split a feather down the centre of the quill, leave 20mm (0.75in) of quill at each end of feather to tie to arrow. It is best to tie three flights equally spaced around the shaft. Arrow heads can be made of tin (B), flint (C), bone (D) or the wood burnt black (E). Do not forget to put a notch in the end of each arrow to fit over your bowstring.

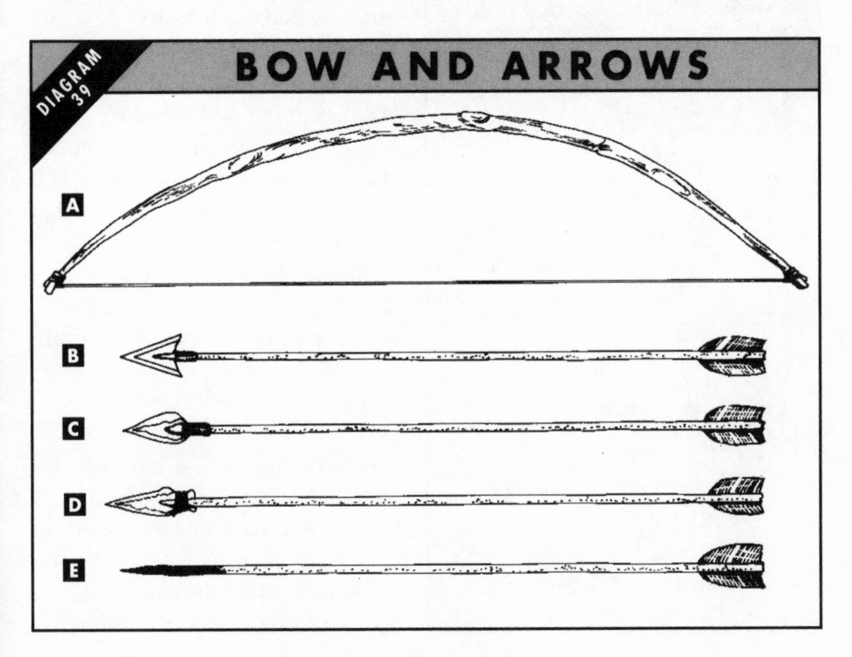

To shoot a bow, place an arrow in the bowstring and raise the centre of the bow to the level of your eyes. Hold the bow in the centre of the stave with your left hand (if you are right-handed) and rest the arrow on top of your hand. Keep this arm locked as you pull the bowstring back with your other hand. Keep the arrow at eye level while you are doing this. Line up the target with the arrow and release the string. Do not snatch at the string as you release it. If you are hunting, it helps to have several arrows with you. Try to carry them in some sort of quiver and try to keep them dry.

Cutlery (Diagram 40) Ladles and spoons can be carved from a large piece of wood where a branch attaches (A). The branch becomes the handle. Look for unusually shaped branches with shapes you can make use of (C). Where the outline of the object follows the grain, this helps to make it more water resistant. A simple fork can be shaped from a piece of stick (B).

Carving is an art in itself, but you should endeavour to learn it if you have the time. Not only will it provide you with useful items, it will also provide you with a way of passing idle time and will therefore keep up your morale, especially at night when you may be at a loose end.

Every wood has its unique carving qualities, and with practice you will learn which will best suit your purpose. Sycamore is soft and can be easily carved; beech is a hard wood that is unsuitable for beginners; hazel is a pliable, stringy wood that is easy to carve but tends to split easily; ash is hard to carve, but you should use it for tool handles, bows and other weapons; birch is a good carving wood but decays easily; yew is very hard and springy and makes fine bows, spoons and bowls.

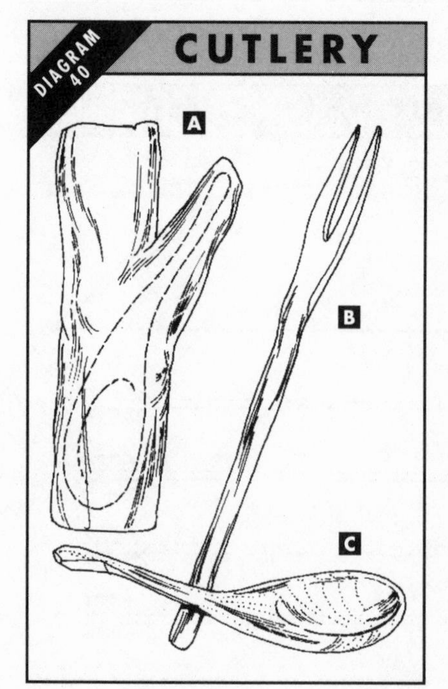

When you are carving, always be aware of the danger of cutting yourself. Never carve towards yourself, when you are tired, with a dull knife, and never carve in a hurry.

Utensils Wood and bark can be useful aids in a survival situation. If you do not want to carve, you can still make utensils from trees and branches. You can use the inner layer of birch bark to construct storage containers or even temporary cooking vessels. The bark can be sewed or tied together.

Bamboo can be fashioned into cups and storage vessels. Cut a section just below a natural joint and then cut below the next joint up. The result is a cup you can drink from, though remember to smooth the edges to prevent splinters from getting into your lips and mouth.

A forked stick can be a major asset to your camp. Drive one into the ground at a 45-degree angle near a fire and rest a longer stick across it. With one end of the stick over the fire and the other driven into the ground and secured with`rocks, you can suspend a pot over the fire. Better still, drive two forked sticks into the ground on each side of the fire, rest a straight stick over them and hang a pot from it over the fire.

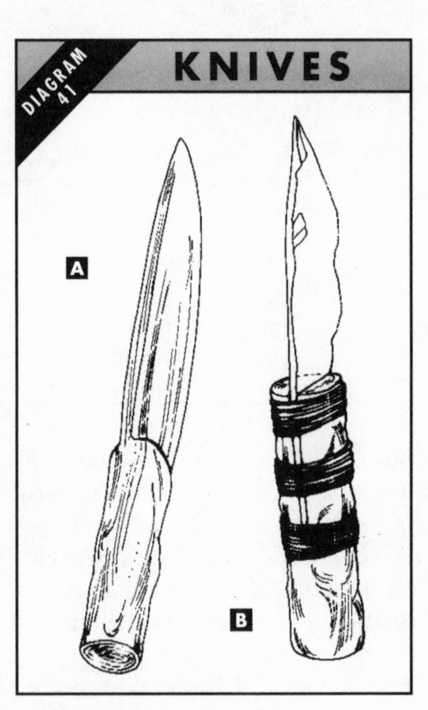

Knives (Diagram 41) Improvised knives can be made from wood (A), bone, stone, metal or even glass. To make a knife from glass (B), simply split a stick, insert a piece of glass and lash securely. To fashion a knife from a piece of bone, sharpen one end (the leg bone of a deer or other medium-sized animal is best) and fashion a handle from the other end. Even the lids of opened tins of food can be driven into a piece of wood and turned into an improvised knife.

US ARMY TIPS

MAKING A METAL KNIFE

A knife is a vital tool to have if you are a survivor. You should have one anyway, but if it breaks or you lose it, make another.

- Find a piece of soft iron which shape resembles a knife blade.
- Place metal on a flat, hard surface and hammer it to get the shape you desire.
- Rub metal on a rough-textured rock to get a cutting edge and point.
- Lash knife onto hardwood handle.

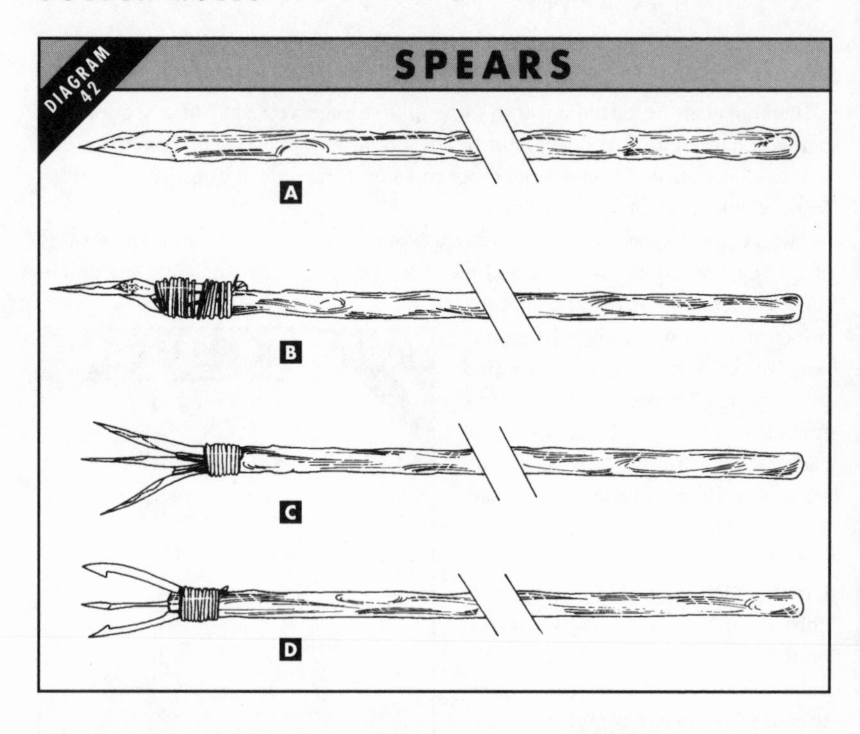

Spears (Diagram 42) Spears can help you catch fish and game. They can be quite simple, e.g. a stick with a sharpened point (A). However, this can break or dull easily. You can also improvise spear heads from antler or animal bone (B), flint or tin. You can split the end of the spear into three pieces and put in spacer blocks to keep them spread apart (C) – this spear can be used to catch fish. Bone can also be used for this purpose (D)

Fish hooks and lines Improvise fish hooks from pins, thorns, a bunch of thorns, nails, bones or wood. You can make a fish snare to trap large fish, such as pike, which often feed alongside weeds. Fix a noose line to the end of a pole or pass a line down the inside of a length of bamboo. To operate pass the noose over the fish from the tail end and pull it up slowly to trap it (for general advice on fishing see Food Chapter).

Rawhide This is an extremely useful material which can be made from any animal hide. It does take a while to process, but the end result is well worth it. Rawhide has many uses, including making lashings, ropes and sheaths for knives and tools.

The first step is to remove all the fat and muscle tissue from the hide (cut off the large pieces and scrape off the rest with a knife or flint). You must then remove the hair, which is done by applying a thick layer of wood ashes to the hair side. Sprinkle some water on the ashes after they have been put on the skin and then roll it up and store it in a cool place for a few days. When the hair begins to slip (you must constantly check to see when it is slipping), the hide should be unrolled and placed over a log. The hair can then be scraped off with a knife or flint.

When all the hair has been removed, you must wash the hide and stretch it inside a frame. Allow it to dry slowly in the shade. Rawhide is very hard when it is dry, but can be softened by soaking in water.

Catapult (Diagram 43A) Select a strong Y-shaped branch and a piece of elastic material (a piece of inner-tube from a tyre is ideal), make a pouch for the centre of the elastic and thread or sew it into position. Tie the ends of each side of the branch. Use stones or small rocks as missiles: with practise you can become very accurate and deadly.

With a sling shot, swing the sling above your head and release one end of the thong to send the ammunition in the direction of the target. When you are using a catapult or sling shot against birds, use several stones at once.

Bola (Diagram 43B) Wrap stones in material and knot each one with lengths of string 90cm (3ft) long. Tie all the lengths together. Hold at the joined end and then twirl around the head. When released the bola covers a wide area. It can be used against birds in flight or on animal's legs, giving you a chance to close in for the kill when the animal is disabled.

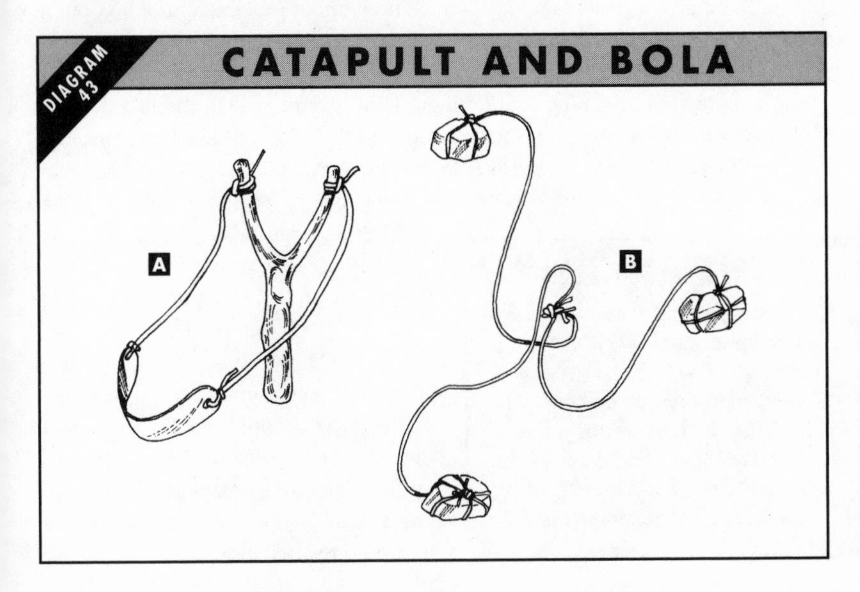

FIRST AID

Wilderness first aid is not complicated, but you must have a thorough knowledge of it so that you can act decisively and quickly when the need arises. Speed is often the key to successful survival medicine. Learn the techniques laid out in this chapter.

Everyone should know how to deal with basic injuries, diseases and disorders. This chapter will provide the survivor with a basic knowledge of first aid. You must learn the techniques listed below. In a survival situation you could be on your own and a long way from civilisation – you will have to rely on yourself to cure any ailments.

PRIORITIES

In any accident situation, where there may be many injured people, always check for dangers to yourself before approaching victims. Watch out for electric cables, fires, gas pipes, falling debris, dangerous structures or wreckage. You should try to give a patient an initial check-up without moving him or her, but if there are dangers you will have to take a chance and move the patient and yourself to a safer location.

If the patient is breathing, ensure there is no obstruction in the mouth, deal with any serious bleeding and place patient in the recovery position (Diagram 44), *THOUGH NOT IF SHE OR HE HAS A SPINAL INJURY* (see below). If patient is lying on back, gently turn him or her on one side by grasping clothing at the hip. In this position any liquids or vomit from the stomach or nose will not block the lungs, and the tongue will not fall back and block the airway.

DO NOT place a patient with a suspected spinal injury in the recovery position: you could cause permanent disability, even death. Instead, use an artificial airway to maintain respiration and as a means of administering mouth-to-mouth resuscitation.

FIRST AID KITS

It is very important to have a good first aid kit. You should carry it with you at all times. Injuries in a survival situation must be taken care of immediately and it will give you an advantage if you have medicines to hand. Do not skimp when it comes to buying a first aid kit: you could be gambling with your life.

BRITISH SAS TIPS

PRIORITIES FOR TREATING PATIENTS

These are the priorities, as laid down by the British SAS, for dealing with injuries in an emergency situation. Do not deviate from this list. Knowing who to treat first can save lives. Prioritise treatment in the following order:

- Restore and maintain breathing and heartbeat.
- Stop bleeding.
- Protect wounds and burns.
- Immobilise fractures.
- Treat for shock.

Note: if a victim has multiple injuries, your treatment priorities for this patient are breathing, heartbeat and bleeding.

With an artificial airway, insert airway up to one third of its length, the end pointing to the roof of the mouth, then turn it through 180 degrees to point down the throat. The patient will be able to breathe. However, mucus build-up may set off coughing and gurgling – you must keep a watch on the patient. If mucus does start to build up, clear it with a mucus extractor. If you don't have one, use a tube or straw.

RESPIRATION/MAINTAINING BREATHING

Every survivor must have a good knowledge of cardiopulmonary resuscitation (CPR), pulmonary resuscitation and opening airways. The instant a person stops breathing and the heart stops beating, he or she is considered clinically dead. Within 4-6 minutes from that time brain damage begins. Some 10 minutes after the heart has stopped there is significant brain cell death. This is called biological death and cannot be reversed. However, clinical death can be reversed in many cases. Breathing stops because of:

DIAGRAM 44

RECOVERY POSITION

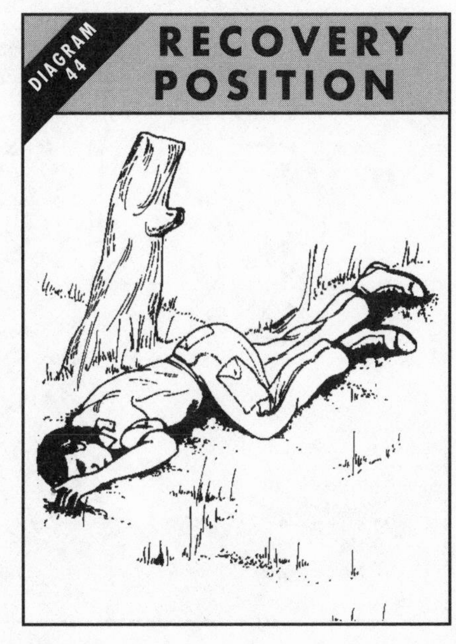

- ☐ Blockage of the upper air passages caused by face and neck injuries or foreign bodies.
- ☐ Choking.
- ☐ Inflammation and spasm of air passages caused by inhaling smoke, gases or flame.
- ☐ Drowning or electrical shock.
- ☐ Compression of the chest.
- ☐ Lack of oxygen.

If a patient's breathing stops you must take action to restart it. Begin artificial respiration at once.

Mouth-to-mouth resuscitation (Diagram 45) Roll patient onto his or her back. Next, open the airway and check for breathing. You do this by using the head tilt: place the palm and fingers of your hand on the patient's forehead and apply firm, gentle backward pressure. This tilts the head backwards and opens the mouth and hopefully the airway. Two other methods of opening the airway are to place one hand on the forehead and the fingers of the other under the chin, or to place one hand on the patient's forehead and the other under the neck.

These three methods may aggravate a spinal injury. If you suspect a damaged spine do not use them, use the jaw thrust: rest your elbows on the ground and place one hand on each side of the patient's jaw. Following the contour of the jaw, push the jaw forward and apply most of the pressure with the index fingers: this will open the airway.

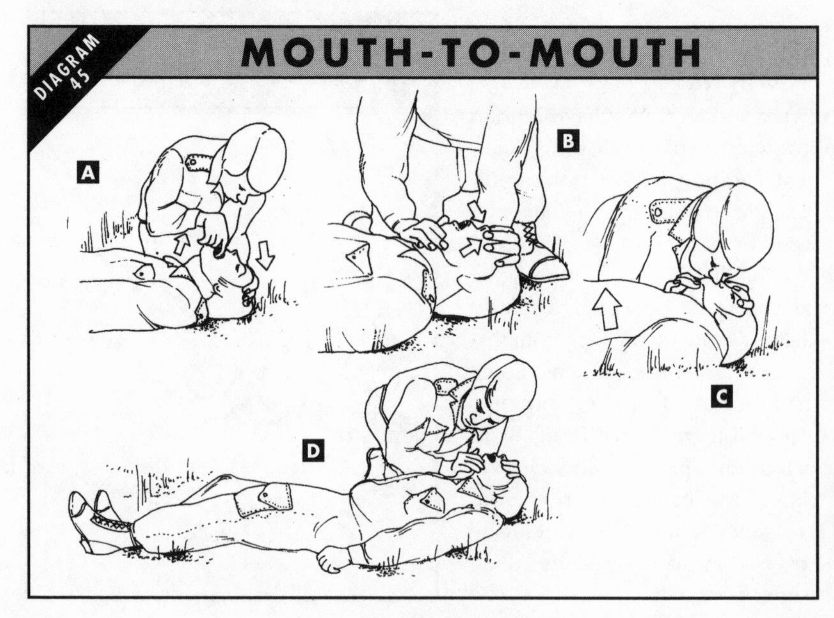

Next, clear any obstruction from the mouth or throat (A). Then place your ear near the patient's mouth to listen and feel for breathing, while also observing the chest for breathing movements.

If a patient is not breathing, give four full breaths quickly by pinching the nose closed (B) and blowing directly into the mouth (make sure your mouth makes a tight seal). These breaths should be quick enough to prevent the lungs deflating between the next breath (C). Check chest falls automatically (D).

Watch for the patient's chest to rise as you blow gently into his or her lungs (if it doesn't rise, there may be an obstruction – turn patient on his or her side and thump between the shoulder blades to remove it).

Repeat as quickly as possible for the first six inflations, then at 12 per minute until breathing is established (be prepared for a long session, and don't give up!). Never blow into the mouth of a person who is already breathing.

Mouth-to-nose resuscitation Use this method if you cannot give mouth-to-mouth resuscitation (if the patient has severe lacerations to the jaw, lips or mouth, or if the jaws are clasped shut). Close and cover the patient's mouth with one hand and blow into the nose. Follow the same sequence as for mouth-to-mouth resuscitation (see above).

Obstructed airway If you cannot get any air into the the lungs, the airway may be blocked. Remove any visible obstruction with your fingers. If you cannot see any obstruction, you must clear the airway by following the techniques as for choking (see below).

Cardiopulmonary resuscitation (CPR) (Diagram 46) This is necessary if the patient's heart fails to function. In this case not only will you have to breathe for the person, but you will also do chest compressions to force the heart to circulate the blood through the body. This can restart the heart beating.

Lie patient on the ground, chest up (A). Trace the edge of his or her ribs with your index and middle fingers until you find the notch at the centre of the lower chest where the ribs meet the bottom of the sternum (B). Keep your index finger on this spot and measure up two finger widths from the notch (C). Try to be as accurate as possible. Place the heel of your other hand just above and touching your fingers that are measuring up. The heel of your hand is now over the patient's heart. *TAKE CARE, YOU CAN DAMAGE THE LIVER BY GIVING COMPRESSION TOO LOW ON THE CHEST.*

Once the heel of your first hand is in position, place your other hand over it. You must be kneeling to perform CPR, with your shoulders directly over the patient's sternum to push straight down on the heart (D). Compress the patient's chest 4cm (2in), with the movement being smooth, strong and rhythmic, never jerky. Give CPR at a rate of 80 compressions a minute (100 for children, 100-120

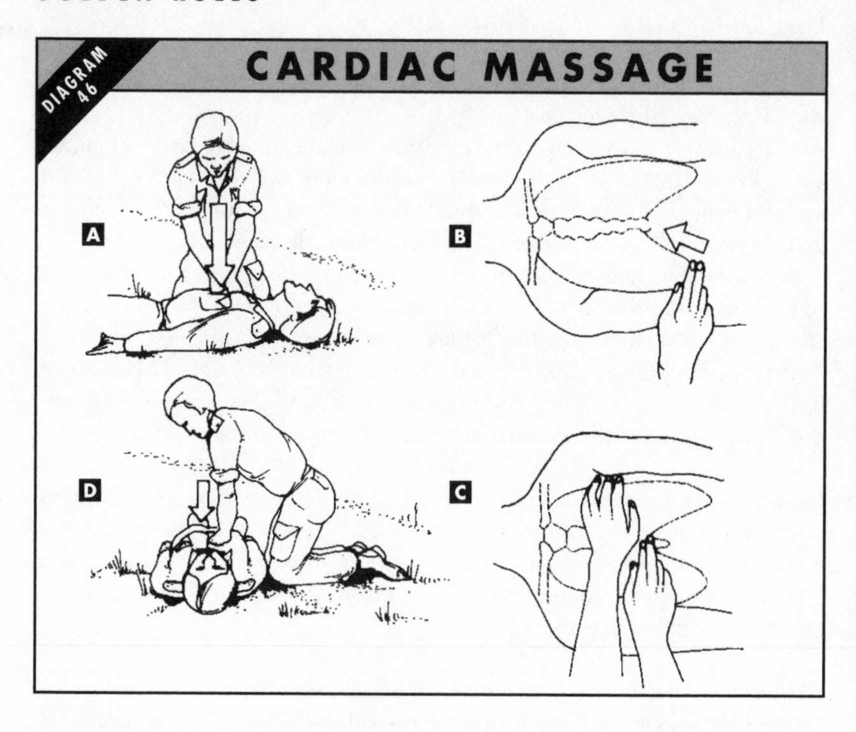

for infants). It helps to count out loud, saying 'one-and, two-and, three-and,...' *DO NOT GIVE UP, CONTINUE FOR AT LEAST AN HOUR IF NECESSARY. TAKE TURNS IF THERE IS A GROUP OF YOU.*

Two-man CPR One person ventilates at the rate of one breath every five compressions. The second person gives compressions at the rate of 60 per minute, counting out loud 'one-one thousand, two-one thousand, three-one thousand, four-one thousand, five-one thousand, one-one thousand', etc. Every one or two minutes, the person ventilating tells the compressor to stop for a pulse check. The ventilator checks for a pulse for a few compressions while the compressor is still giving compressions to ensure that the compressions are creating adequate circulation, then will say 'stop for pulse check', and will check for pulse and look and listen for any breathing.

If there is no pulse, the ventilator gives a full breath and says 'continue compressions'. If there is a pulse but no breathing, the ventilator says 'we have a pulse' and continues ventilating at one breath every five seconds. If there is a pulse and breathing, any other treatment the patient requires can now be given. With a two-man CPR, you must be aware of the following:

☐ The person giving ventilations must time himself to the compressor's count.
☐ The ventilator should take in a deep breath on the count of four and blow into

CANADIAN SPECIAL FORCES TIPS

SEQUENCE FOR CPR

Timely CPR can save a patient who would otherwise die. Learn the following Canadian Special Forces checklist for administering CPR.

- Check for consciousness.
- Establish and open airway.
- Look, listen and feel for breathing.
- Give four rapid breaths.
- Check for pulse (while you look, listen and feel for breaths).
- Locate the compression spot.
- Form proper hand position.
- Begin compressions: set of 15 compressions, then...
- Two quick breaths after each set of 15 compressions.
- After four complete sets of 15 compressions, two breaths, check for at least five seconds for pulse and breathing.

the patient exactly at the end of the fifth compression, just when the chest is beginning to rise.

☐ When changing places, the ventilator will give one more breath after the fifth compression and move into position to begin compressions. The second person, meanwhile, has begun a five-second pulse and breathing check. If there is no pulse or breathing, the ventilator gives a breath and the sequence continues.

CHOKING

The following signs indicate that a person is choking:

☐ Patient is holding his or her throat.
☐ Inability to speak.
☐ Wheezing sounds and an effort in breathing.
☐ Inability to forcibly cough.
☐ Skin appears blue (in an unconscious patient).
☐ Chest not rising (in an unconscious patient).

When a patient is choking (Diagram 47), clear the airway with a finger and make sure the tongue has not fallen back and is obstructing the breathing passages (see above). Then, administer four back blows (A), ensuring patient's head is lower than his or her chest. Strike the patient over the spine between the shoulder blades quickly and forcibly, but remember you are not trying to break his or her back: be sensible!

If this fails, try the Heimlich manoeuvre. Stand or kneel behind the patient with your arms around him or her (B). Clench one hand over the other, thumb side of fist pressing between waist and bottom of ribs (C). Apply pressure and

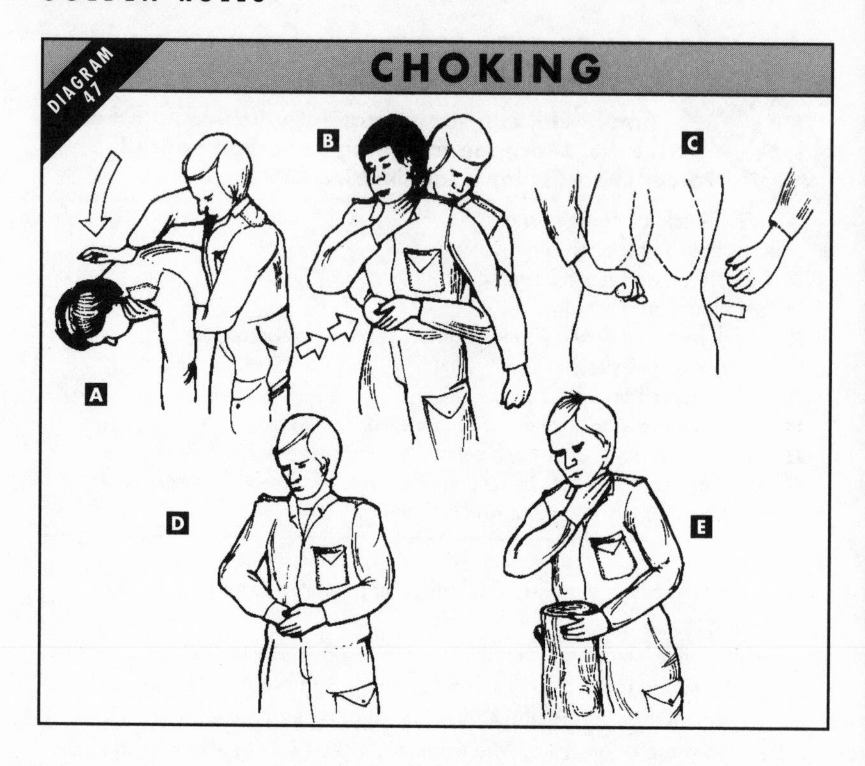

jerk quickly upwards four times. If this fails, go back to four back blows. Repeat process until airway is clear or patient loses consciousness.

If patient loses consciousness, lay him or her on his side and give four back blows. If unsuccessful, roll patient on side and give four abdominal thrusts using the heel of your hand. After four thrusts, grab hold of patient's lower jaw and tongue and pull the jaw out and upward to open the mouth. Sweep the mouth with your index finger, then tip the head into the CPR position (see above).

If there is no breathing, give four quick breaths. If lungs will not inflate, repeat sequence of four back blows, four abdominal thrusts, mouth sweep, check for breathing and four quick breaths. If there is no pulse, begin CPR.

If you are alone and are choking, give yourself abdominal thrusts with your hands (D) or use a blunt projection – earth bank, fallen tree (E) or chair back.

Cricothyrotomy (Diagram 48) Should only be used when ALL procedures to clear an obstructed airway have failed. The operation is simple but should be considered dangerous. Sterilise the area around the Adam's apple and sterilise your tools (blade, scalpel or penknife and a hollow tube).

Locate the cricothyroid membrane between the Adam's apple (thyroid cartilage) and the cricoid cartilage. Do this by running a finger down the Adam's apple and finding another small projection just below it. Between the Adam's apple and this smaller projection is a central valley (A).

At the midpoint make a small but deep incision. 1-2cm (0.6-0.75in). Do not push down further. Insert the tube and push it down to keep the cut open and air into the lungs. Secure firmly in place with tape or bandage (B).

Though this operation is simple to perform, you risk cutting the recurrant laryngeal nerve, the esophogus or even the upper part of a lung. It is therefore very important that you are sure of your location before you cut.

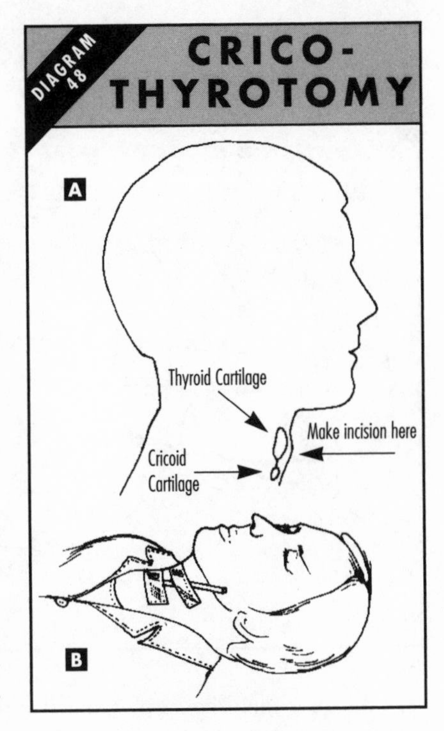

BREAKS AND FRACTURES

A fracture is a chip, crack or break of a bone. A stress fracture is a tear of the bone tissue. There are two types of fracture: open and closed.

With an open fracture, the bone has come through the skin or something has penetrated the skin and then broken the bone. With a closed fracture, the bone is broken but there is no opening of the skin.

US SPECIAL FORCES TIPS

LIST OF SYMPTOMS TO IDENTIFY FRACTURES

The Green Berets are experts in survival medicine. Use their training and learn to identify the symptoms that can indicate a fractured limb.

- Patient feels or hears the bone break.
- Partial or complete loss of motion.
- Grating sound when limbs are moved.
- Deformity and abnormal motion at fracture site, such as arm bending but not at the elbow.
- Tenderness around the injury.
- Muscle spasm.

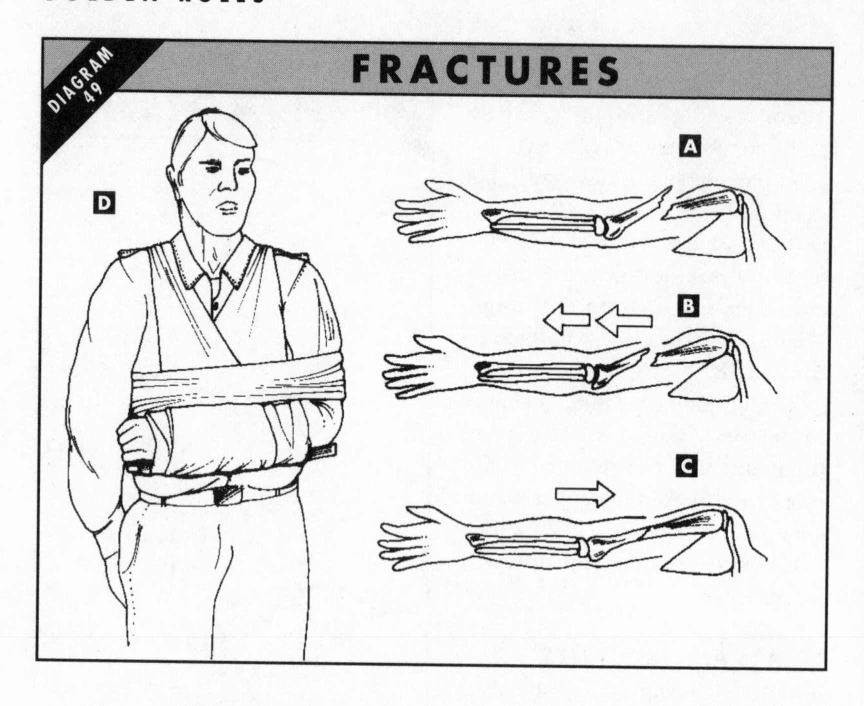

A fractured limb should be splinted in place in exactly the same position you find it in to prevent further damage while the patient is being taken to hospital. Use padding to keep the limb in position.

Closed fracture If you cannot get to a hospital, you will have to administer treatment yourself. If you suspect a closed fracture, check for pulse at the wrist. If circulation is impeded (the hands feel cold or there is no pulse) you must restore the flow of blood to the lower arm at once, otherwise the limb will have to be amputated.

Pinch or poke the hand for any sensation. If there is no feeling the fracture should be considered very serious. Apply traction (a continuous pull) and try to restore pulse and nerve response. Don't worry if you only have partial nerve response, you will have to wait until you get back to civilisation for complete restoration (full nerve response can often be restored by surgical techniques – reassure the patient of this).

Now release traction slowly, checking for pulse and nerve response. If they remain, splint securely (immobilise the joint on either side of the fracture). If there is internal bleeding (there is extreme discolouration) but no bone chips, apply moderate pressure and elevation to control the bleeding. If you think there are bone chips, use the pressure point (see below) nearest to the fracture, elevate limb and apply cold packs.

US SPECIAL FORCES TIPS

TREATING SPINAL INJURIES

Spinal injuries are extremely dangerous. Always be careful when attempting to move a patient with a spinal injury.

- If patient is face-up, place a folded blanket under small of the back to stop bone fragments lacerating or compressing the spinal column.
- If patient is face-down, place a folded blanket under the chest.
- Always move the entire spinal column as a non-flexible unit.
- Use a rigid litter or board longer than the patient is tall for transportation.

Open fracture (Diagram 49) First, control bleeding if it is severe (see wounds and dressings). Check for nerve response then irrigate the wound and the protruding end of the bone for any bone chips or foreign bodies (A). Remove.

Then traction and try to re-set the fracture and close the wound. Do it slowly (B). After traction, inspect the wound to ensure the bone ends align properly (C). Then splint (D), but in a way that allows you to work on the wound (always check for pulse and nerve response, if there aren't any you must repeat re-setting and traction procedures).

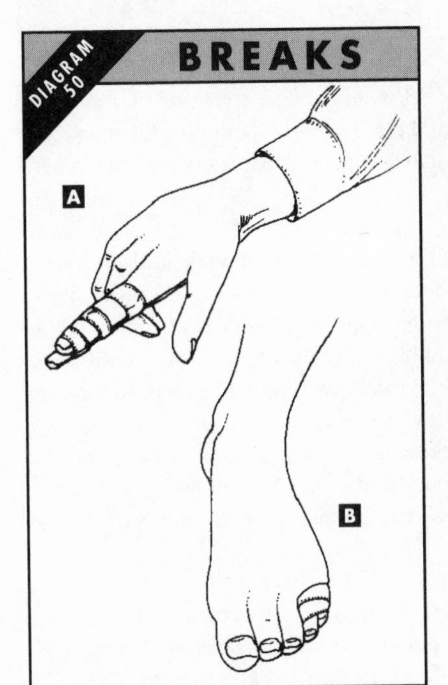

Breaks (Diagram 50) In a survival situation it is usually toes and fingers that are broken. A finger should be re-set (see p.119) and splinted with wood or similar (A). Broken toes should be re-set and taped to an unbroken toe next to it (B).

Spinal injuries Any injury to the spinal column can cause paralysis and is potentially fatal. Signs of spinal injury are: pain in the back without movement; deformity in the spinal column; any spot along the spinal column tender to the touch; persistent erection of the penis; arms uncontrollably extended above the head; loss of bladder control. See the box on this page for how to treat spinal injuries.

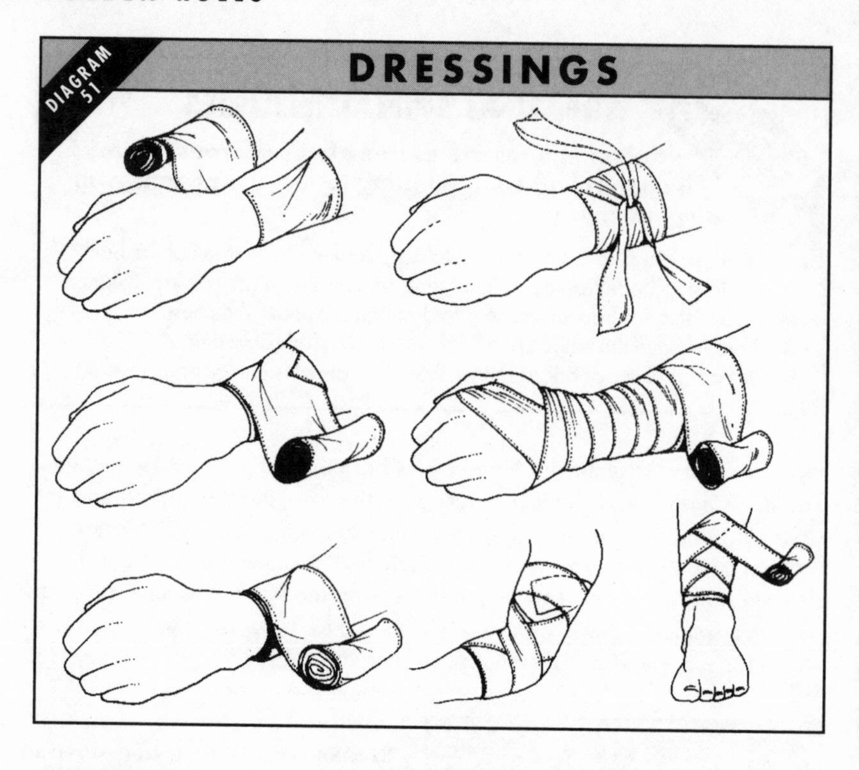

Neck fracture Immobilise patient's neck with cervical collar, or place a small rolled towel or sheet under the neck and place sandbags or boots filled with dirt or sand on either side of the head to stabilise it. Keep patient completely immobilised and hope for speedy rescue.

Fractured ribs For fracture of the upper ribs, have patient hold his or her breath while you apply two long adhesive strips across the shoulder of the injured side. For fracture of the lower ribs, apply a piece of felt or foam rubber over the fracture. Have the patient hold his or her breath while you apply adhesive strips around the injured side of the trunk. An alternative method for treating fractures of both the upper and lower ribs is to apply an elastic bandage around the trunk of the body from below the costal cage to just below the level of the nipples. Whatever treatment you use, the fracture will take 4-6 weeks to heal. Fractured ribs can be painful; it is therefore important for the patient to get as much rest as possible.

Skull fracture Indications of a fractured skull can be straw-coloured fluid seeping from the ear or nose. Place patient in recovery position, leaking side down. Allow fluid to escape, completely immobilise and keep comfortable.

BRITISH SAS TIPS

DEALING WITH WOUNDS

It is vital to prevent a wound from becoming infected. The SAS advises the following for the care of wounds in the wild:

- Clean a wound from the centre outwards.
- Change dressings if they become wet, omit offensive smells or if the pain in the wound increases and throbs, indicating infection.
- Local infections can be treated with a poultice. Anything that can be mashed – rice, bark, seeds – can be used. Boil and wrap in a cloth. Apply to infected area as hot as can be tolerated.
- A warm rock wrapped in cloth and applied to wound can aid healing.

BLEEDING, WOUNDS AND DRESSINGS

When a patient is bleeding heavily, you must take immediate action to stop it. If patient is bleeding from the veins or capillaries, apply pressure over the bleeding point (even minor arterial bleeding can be controlled with local pressure). Bleeding extremities should be elevated above the heart. Use anything to staunch the blood flow, but ensure it is clean. Maintain a firm, continuous pressure for 5-10 minutes. Use a dressing (Diagram 51) to keep the wound clean.

Arterial bleeding This is the most serious type of bleeding. Pressure on the major artery feeding an injured area will reduce or stop the flow of blood. The spots on the body where an artery can be easily compressed against a bone are called pressure points (see box below and Diagram 52E).

Tourniquets (Diagram 52) Use when severe bleeding cannot be controlled by any other method. You can only place them on the upper arm (just below the armpit) and around the upper thigh. Use a cloth AT LEAST 5cm (2in) wide (A). Wrap the cloth around limb and tie a half knot (B). Place a stick over the knot and tie a double knot over it (C). Twist stick to tighten tourniquet until bleeding stops (D).

BE WARNED, the loosening of a tourniquet can cause severe shock leading to death (toxins build up in the

PRESSURE POINTS FOR ARTERIAL BLEEDING

- Temple, forward of ear.
- Face below eyes, side of jaw.
- Shoulder or upper arm, above clavicle.
- Elbow, underside of arm.
- Lower arm, crook of elbow.
- Hand, front of wrist.
- Thigh, midway on groin/top of thigh.
- Lower leg, upper side of knee.
- Foot, front of ankle.

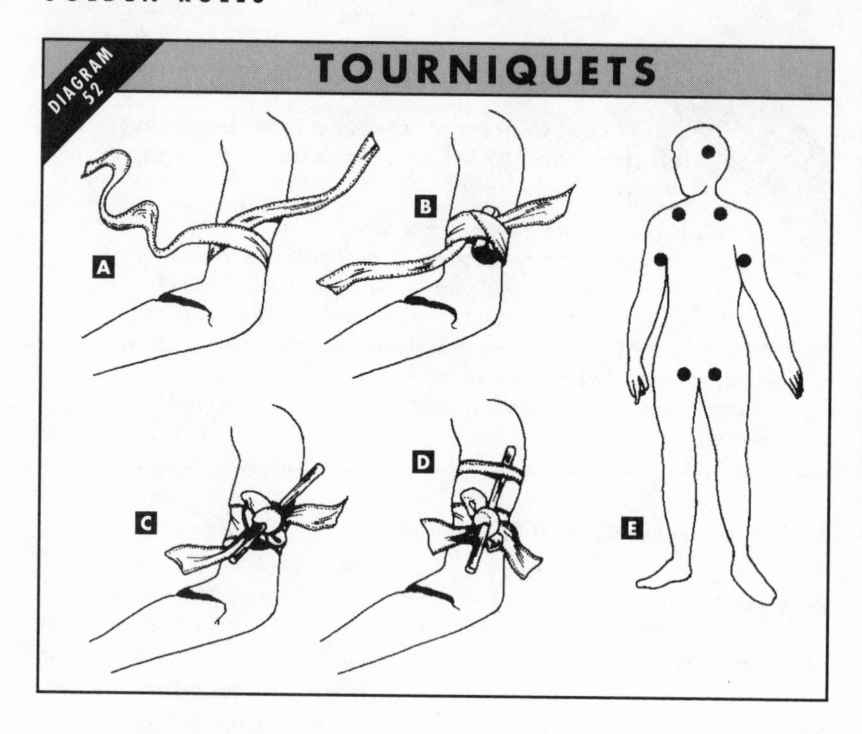

injured limb which flood the heart when released). If a tourniquet is applied, it must be assumed the limb will be eventually amputated.

Internal bleeding Results after a violent blow, broken bones or deep wounds. Internal bleeding is indicated by faintness, light-headedness, pale skin which is cold and clammy to touch, red-coloured urine, blood passed with faeces, faeces with black, tarry appearance, vomiting blood and coughing up blood.

BRITISH SAS TIPS

TREATMENT FOR SPRAINS

SAS soldiers fight on foot most of the time, and can suffer sprained ankles. They therefore have to have effective treatments for sprains:

- Bathe sprain with cold water to reduce swelling.
- Support with bandage: do not constrict circulation.
- Elevate affected limb and rest completely.
- If you sprain an ankle, keep your boot on if you have to keep walking: the boot will act as a splint. If you take it off the swelling will prevent you putting it back on again.

To treat, lie patient flat on his or her back with legs elevated. Keep warm and hope for early rescue.

Wounds Irrigate (wash by pouring or squirting, not scrubbing) with sterile saline solution or clean water. Close the wound (use butterfly bandages if not too deep). This is where your medical kit comes in handy, as you should coat the wound lightly with an antibiotic ointment and then apply a sterile dressing and bandage.

Amputation If, for any reason, a part of your body has been torn off, you must stop the bleeding and treat for shock (see below). Irrigate and disinfect the area, apply antibiotic ointment and cover with a dressing. Place a sterile gauze pad over the dressing and bandage securely. Repeat this process each day. Each day scrub away all the yellowish crust and dead tissue.

If you have to amputate a limb, cut as close to the wound site as possible (you will need a saw). Sever at the nearest joint if you do not have a saw. Apply a tourniquet, make an incision in the skin and into the underlying tissue. Allow skin to retreat, then saw through the bone or joint. Tie off arteries (see below) and apply a light bandage to protect the stump.

Suturing (Diagram 53) Used when the wound is deep and butterfly sutures will not work. When a wound is closed, there must be no pockets of air or blood left

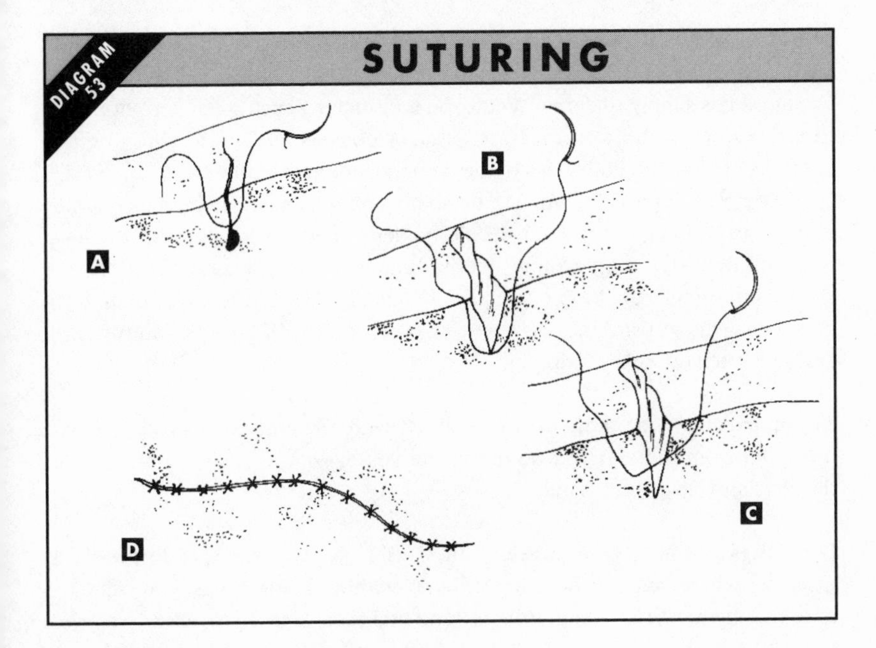

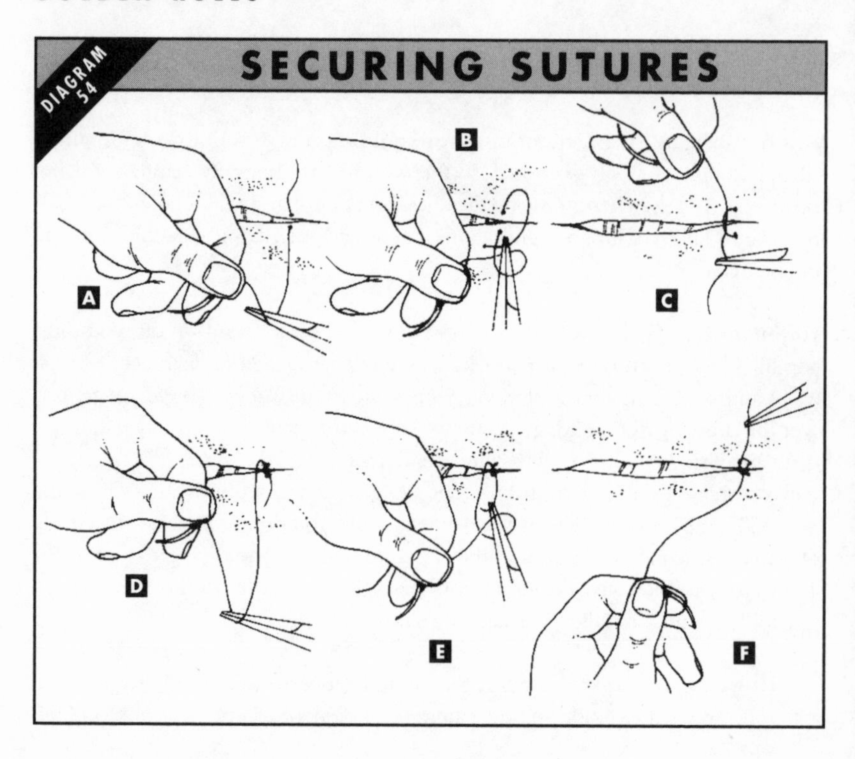

below the skin (A). You need one needle holder (forceps with a needle groove) and suture material.

Suturing is simply stitching. When you are stitching each stitch you should go to the bottom of the wound (B) to prevent pockets (C) and to take in equal amounts of tissue on both sides of the wound to align the edges (D).

To secure sutures (Diagram 54), tie sutures off with a square knot (see Ropes and Knots Chapter). Loop over needle holder (A), grasp end through the loop (B), pull tight (C), loop around needle holder in the opposite direction (D), grasp end through the loop (E), and pull tight (F). Leave sutures in for 10 days and then remove with fine scissors. Grasp the knot with forceps and tweezers and pull stitch out with a firm pull.

Tying arteries Clamp the blood vessel off with a haemostat (locking forceps) upstream from the end and tie the blood vessel with a square knot. Irrigate, disinfect and close the wound.

Dressings Do not re-use dressings, they act as poison collectors. In a survival situation, where you may be using strips of clothing for dressings, you will have to re-use them. Wash them thoroughly and boil for at least 15 minutes.

STRAINS, SPRAINS AND DISLOCATIONS

All can be common in a survival situation. A strain is a tearing or over-stretching of a muscle. A sprain is a wrenching or tearing of tissues connected with the joint. A dislocation, usually caused by a fall, blow or sudden force applied to a joint, forces the joint out of place.

For strains, rest the limb and apply cold packs to ease the pain. Apply to the area straight after injury to reduce the swelling and pain. Immobilise and treat like a fracture (see above).

Dislocations (Diagram 55) When dislocations occur, swelling will begin and the injury will be very painful (A & B). The joint must be re-set before the swelling and muscle spasms make re-setting difficult (the muscles near the joint will start to tighten up almost immediately). If you fail to do this, you risk the patient getting gangrene or a permanent deformity.

There are two things you should remember when re-setting a joint:

- Do it properly.
- Do it as soon as possible.

To re-set, traction the joint and then move the extremity attached to it in the direction that it would normally move. This should re-align the joint and take pressure off blood vessels and nerves (C). Release traction and check for nerve response. If there is a pinched nerve repeat procedure. Application of cold packs will help reduce pain and swelling. Rest the extremity until fully healed.

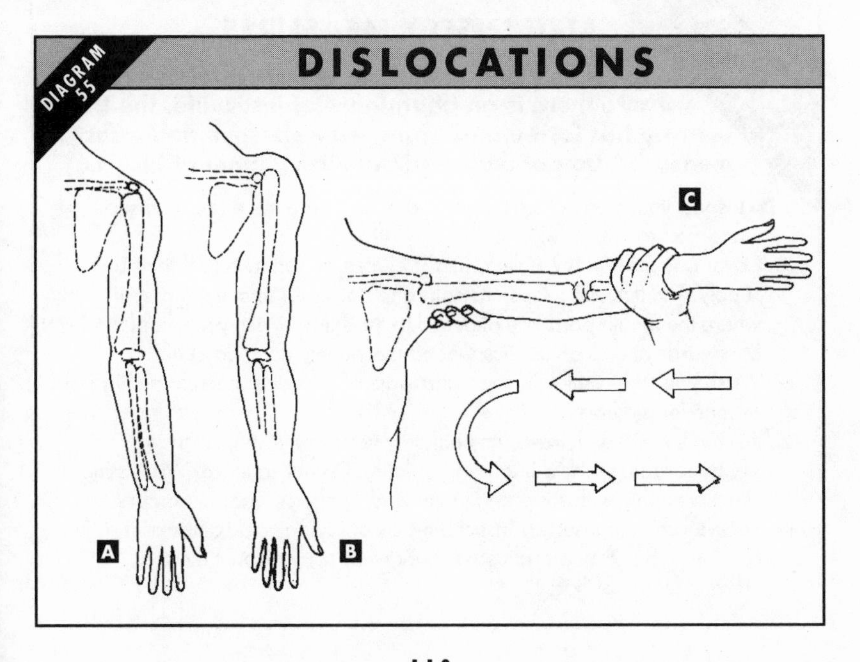

DISEASES

Specific diseases will be dealt with in later chapters. However, you should be aware of the general dangers you face in a survival situation. The main danger will come from disease-carrying insects:

- □ Lice can transmit typhus and relapsing fever.
- □ Mosquitoes can carry malaria, dengue fever and other diseases.
- □ Flies can spread diseases, such as sleeping sickness, typhoid, cholera and dysentery.
- □ Fleas can transmit plague.
- □ Ticks can transmit Rocky Mountain spotted fever.
- □ Bee and wasp stings can be fatal to people who are sensitive to their venom.

POISONING

The main kinds of poisoning a survivor will face are eating dangerous plants, animals and marine life (see Food Chapter) and being bitten or stung. Beware of bites and stings from spiders, centipedes, scorpions and ants. They can be very painful and can make you ill, even kill you (try to learn what creatures inhabit the area you are in) . To treat a bite from a scorpion or spider, clean the wound and try to remove the poison by suction or squeezing the bite site (though this may prove fruitless). If you have tobacco, chew it and place it over the bite site to ease the pain. Treat the bite as an open wound (see above).

US ARMY TIPS

ANTI-INSECT MEASURES

As its soldiers often have to fight in regions where there is an abundance of insect life, the US Army has formulated some very effective anti-insect measures. Do not underestimate the danger of insects.

- ■ Inspect your body at least once a day to ensure there are no insects attached to you.
- ■ Cover any ticks with vaseline, heavy oil or tree sap to cut off their air supply. The tick will release its hold and you can remove it (grasp it where the mouth parts are attached to the skin). Wash your hands afterwards and clean the tick wound thoroughly each day.
- ■ Wash your skin well with soap and water if you have been in a chigger- or mite-infested area.
- ■ If stung by a bee or wasp, immediately remove the stinger and venom sac by scraping with a fingernail or knife. Do not squeeze. Wash sting site thoroughly with soap and water and apply ice pack or compress.
- ■ Relieve itching caused by insect bites by applying cold compresses, a cooling paste of mud and ashes, dandelion sap, coconut meat or crushed leaves of garlic.

BRITISH SAS TIPS

TREATMENT FOR POISONING

SAS soldiers have fought in the exotic regions of the world for over 50 years, and they have well-tested rules for dealing with poisoning.

- With suspected plant poisoning, induce vomiting.
- Alternatively, make an antidote: mix tea and charcoal with an equal part of milk of magnesia if available. The charcoal absorbs the poison and carries it from the body.
- Wash poisoned skin with soap and water and remove contaminated clothing.
- Sluice chemical poisons off the skin with water (though try to ascertain the properties of the chemicals to which you have been exposed).
- With inhaled poisons, move the patient to fresh air, loosen tight clothing and give artificial respiration.

SNAKE BITE

Despite wild rumours to the contrary, the chances of being bitten by a venomous snake in a survival situation are small. Nevertheless, it can happen and you must know how to treat snake bites.

Snake venom is divided into two categories: hemotoxic and neurotoxic. Many venomous snakes inject poison that contains both types of venom. Snakes themselves are split into a number of types:

- ☐ Crotalidae – pit vipers.
- ☐ Elapidae – coral snakes, kraits, cobras, mambas and asps.
- ☐ Hydrophine – sea snakes.
- ☐ Colubridae – backfanged boomslang.

Symptoms of snake bite

CROTALIDAE: tissue swelling at the bite site, gradually spreading to surrounding area. Swelling begins within three minutes and may continue for an hour. There is severe pain at the bite site, fang marks, bleeding from major organs that can show up as blood in the urine, destruction of blood cells and other tissue cells. The victim will suffer severe headaches and thirst, a fall in blood pressure and a corresponding rise of pulse rate, and bleeding into surrounding tissues. Death can occur within 24-48 hours if bite is serious and untreated, and there is a real danger of a loss of limbs (vipers account for the majority of snake bite fatalities throughout the world).

ELAPIDAE AND COLUBRIDAE: irregular heartbeat, drop in blood pressure, weakness and exhaustion, severe headaches, dizziness, blurred vision, confusion, muscular uncoordination and twitching, respiratory difficulty, tingling,

US SPECIAL FORCES TIPS

SNAKE BITE TREATMENT

The Green Berets have devised the following treatment for snake bites in the wild:

- Kill the snake if you are the one bitten: it will make you feel better and makes species identification easier.
- Lie patient down and immediately immobilise injured part.
- Keep patient warm and quiet.
- Begin specific treatment for snake types:

Crotalidae:

- Make an incision 0.31-0.63cm (0.125-0.25in) deep along, or in the direction of, the muscles through the puncture site. Do not make an X-cut and do not cut into joints.
- Apply suction using a mechanical device; use mouth only as a last resort, and only if you have no cavities, cuts or sores in the mouth.
- Incision and suction should NOT be used if antivenom can be given within one hour or if one hour or more has elapsed since the bite.
- Do not use a tourniquet, constricting bandages or cold packs.
- Do not allow patient to eat food or drink alcohol.
- Give patient small amounts of water at frequent intervals.
- Initiate IV DSW saline to help prevent shock (if available).
- Administer antivenom.
- Use morphine or other suitable pain killers as necessary.

Elapidae and colubridae:

- Apply a tourniquet around the affected limb over a single bone (above the knee or below the elbow) approximate to the bite, and tight enough to stop arterial flow. It should be released for 30 seconds every 20 minutes to allow fresh blood into the affected area.
- Administer antivenom.
- Do not use morphine or other drugs that cause respiratory depression.

Hydrophine:

- Antivenom is the only treatment, incision and suction are of no value.

excessive perspiration, numbness of the lips and soles of the feet, chills, nausea, vomiting and diarrhoea and unconsciousness.

HYDROPHINE: the bite is usually painless. However, a bite should be suspected 1-2 hours before the onset of muscular aches, pains, and stiffness, pain on passive movement of arm, thigh, neck or trunk muscles, reddish-brown urine within three hours, plus neurotoxic symptoms as described for elapidae bites. Death usually occurs within 12-24 hours without treatment.

All snake bite victims will need treatment for shock. If no antivenom is available, place a restricting bandage, *NOT A TOURNIQUET,* above the bite and bandage down over the site. Use cool water or ice to keep as cool as possible.

BURNS

Burns can be life-threatening in a survival situation. There are three types, which you must know in order to take appropriate action :

- ☐ First degree, which usually involve the first layer of skin. Not serious.
- ☐ Second degree, which involves the second layer of skin. These burns are very red, produce blisters and are intensely painful for up to 48 hours. There is fluid loss and a danger of infection.
- ☐ Third degree, which destroys the first two layers of skin and damages deeper tissues. There is severe fluid loss and a danger of infection. The burned area is usually charred black, and the victim will suffer great pain.

In a survival situation you will have to replace fluids orally. Drink lots of water but keep an eye on your urine flow.Cut down if it becomes very high. Be prepared to treat for shock.

SHOCK

A condition that is caused by the loss of an effective volume of blood circulating in the patient's blood vessels. It can be caused by loss of blood through bleeding, loss of blood into the tissues, such as in a broken thigh, and loss of fluids through sweating, vomiting and diarrhoea. Shock is likely if a person has pale, cold and clammy skin, a fast and weak pulse, and fast and shallow breathing.

To treat, ensure airway is open (see section on respiration and maintaining breathing). Any fractures should be treated (see above) and splinted (Diagram 56). Keep patient warm and still. Any rough handling of patient suffering from shock is very dangerous and therefore forbidden. This sounds obvious, but in your efforts to treat a patient you may be too violent. Try to think.

CANADIAN SPECIAL FORCES TIPS

TREATMENT FOR BURNS

Military units obviously have to deal with explosive and munitions, which can cause horrific burns. Here is how Canada's elite deal with them.

- Irrigate burn immediately with sterile, cool water.
- Remove all foreign matter from burn.
- Irrigate with peroxide followed by a light coating of iodine. Then coat with an antibiotic ointment.
- Cover with a non-adhering sterile gauze pad. DO NOT use an air-tight dressing.
- Remove dressing EVERY day and scrub wound with a sterile gauze pad and peroxide.
- All white and yellow dead tissue must be removed every day.
- Repeat cleansing and dressing process.

NATURAL MEDICINE

Use the plants, herbs and natural substances around you to treat illnesses. Remember, many modern drugs are made from the plants around you. Infusions (an extract obtained by soaking) are usually made from leaves or flowers and decoctions (extracts obtained by boiling) from roots. Administer the doses obtained thrice daily and always use only fresh plants and flowers. Resist the temptation to take larger doses – it can do more harm than good.

Cleaning wounds and sores (apply externally to skin)
INFUSIONS: camomile, cleavers, nettle, elm, horehound (not roots), sanicle (not roots), silverweed (not roots), woundwort (not roots) and yarrow (not roots).
DECOCTIONS: burdock, comfrey, mallow, marsh mallow and oak bark.
JUICE: chickweed and watercress.

Antiseptic (use internally or externally) Garlic juice, mallow leaves and flowers, marsh mallow (decoction), horseradish and thyme.

Fevers Camomile, elder, elm and lime.

Colds and sore throats
INFUSIONS: agrimony, bilberry, bistort, borage, camomile, comfrey, horehound,

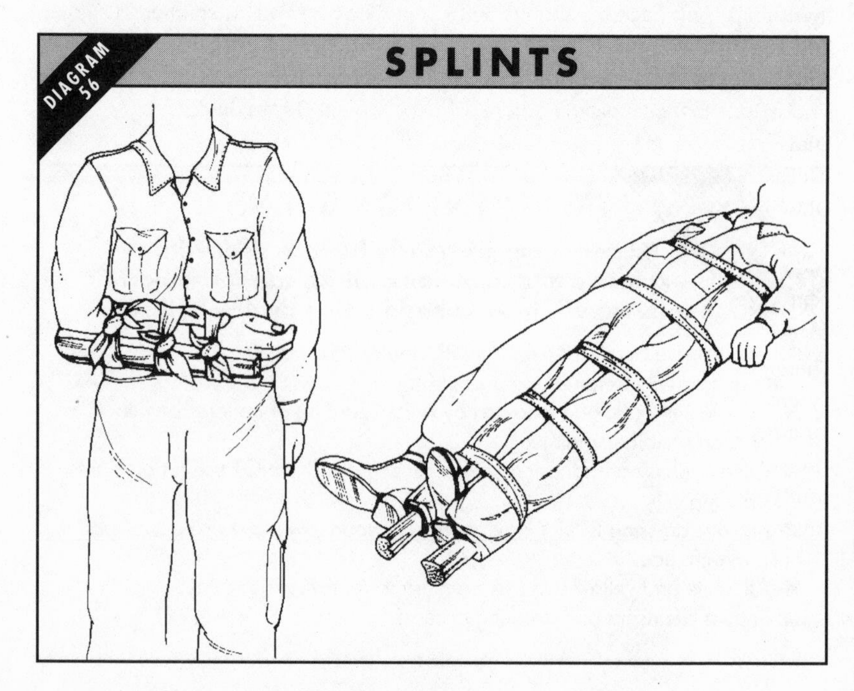

US ARMY TIPS

TREATMENT FOR SHOCK

Shock can be a killer. In the wild you must recognise the symptoms and treat them. Follow the guidelines of the US Army.

- If conscious, place patient on a level surface with the lower extremities raised 15-20cm (6-8in).
- If unconscious, place patient on his or her side or abdomen with head turned to one side to prevent choking.
- Once patient is in shock position, do not move.
- Keep patient warm.
- If patient is wet, remove all wet clothing as quickly as possible and replace with dry items.
- Insulate patient from the ground with clothing, tree boughs etc, and make a shelter to insulate him or her from the weather.
- Use hot liquids, food or body heat to provide external warmth.
- Only administer liquids or foods if patient is conscious, and do not give if patient has abdominal wounds.
- Patient should rest for at least 24 hours.

lime, lungwort, mallow, marsh mallow, mint, nettle leaves, plantain, sanicle, thyme and yarrow.
DECOCTIONS: angelica, burdock, marsh mallow, oak bark, rose hips and willow bark.

Diarrhoea
INFUSIONS: bistort, bramble, elm bark, hazel leaves, marsh mallow, mint.
DECOCTIONS: bilberry fruit, cowberry fruit and oak bark.

De-worming
INFUSIONS: bracken roots, figwort and tansy.

There are some other cures for ailments you should take note of. If you have dysentery, try swallowing charcoal. Alternatively, pull bark from trees and boil it for anything from 12 hours to three days. Keep replacing the evaporating water. The resulting substance will be black and smell bad. However, it will help to cure you. Tea contains tannin, which can also help to cure dysentery. If you have worms, drink a small amount of petrol if you have it. It will make you feel ill, but it will make the worms even more ill!

You can use maggots to clean out a wound. Just expose the wound to the open air and the maggots will appear, especially in a warm climate. Be careful they do not tuck into the good flesh after eating the bad, though.

SIGNALLING AND NAVIGATION

It is imperative that you can navigate accurately if you decide to move from your camp and head for civilisation. In addition, it is important to know how to signal to attract the attention of a search aircraft to your position on the ground.

If you are a backpacker you will have researched the area you will be travelling through beforehand, and will be equipped with maps of the area. Study them: they will tell you about prominent terrain features and will allow you to work out routes. In addition always make a note of the wind direction, first and last light and weather patterns. It will all be useful for determining your position. If you are a survivor you probably will not have a map, so it is doubly important to know as much about the terrain as possible.

Navigation is not just about reading a map: it also concerns being able to read the landscape, telling the direction from any visible landmarks and looking at the land and picturing it as it appears on a map.

MAPS

If you have a map, it is important to have one that fits your requirements. For example, it is no use having a map that has a very large scale and shows every detail of the land if you are travelling over a distance of thousands of kilometres. This may seem obvious, but it is a mistake many people make.

Maps contain a wealth of information. Do not ignore this immense amount of detail at your fingertips. Learn to interpret it so you can make use of it.

CONTOURS: depict relief of the ground. The intervals may be in metres, with the approximate value in feet indicated in the margin.

GRADIENT TINTS: show relief and are supplemented by shaded relief.

SCALE: located in the margin. The scale is expressed as a fraction and gives the ratio of map distance to ground distance.

LEGEND: located in the margin. It illustrates and identifies the topographical symbols used to depict the features on the map. The symbols are not always the same on every map: it is therefore important to refer to the legend of the map you are using.

BAR SCALES: located in the margin. They are rulers used to convert map distance to ground distance. Maps usually have three or more bar scales, each a different unit of measurement, such as miles and kilometres.

CONTOUR INTERVAL: located in the margin. It states the vertical distance between adjacent contour lines on the map.

BLACK TOPOGRAPHIC SYMBOLS: denote man-made features, such as roads, buildings and pipelines. They are also used to denote rock features.

BLUE TOPOGRAPHIC SYMBOLS: denote water features, such as lakes, oceans, rivers and swamps.

GREEN TOPOGRAPHIC SYMBOLS: denote vegetation, such as woods, forests and vineyards.

BROWN TOPOGRAPHIC SYMBOLS: denote all relief features, such as contours.

RED TOPOGRAPHIC SYMBOLS: denote main roads.

DARK BLUE TOPOGRAPHIC SYMBOLS: denote motorways.

YELLOW TOPOGRAPHIC SYMBOLS: denote minor roads.

THE THREE NORTHS

The north represented by the grid lines on your map may differ from the north from which you are gaining you physical orientation.

True north: the celestial north which is gained from accurate sun readings or from the stars.

Grid north: the north with which map grid lines are in alignment, and from which map bearings are taken.

Magnetic north: the north to which a compass points, and from which all magnetic land bearings are taken.

You must be of aware of these variations to take accurate bearings. If you have an adjustable compass and know the extent to which it and your map deviate from true north, you can match them all up to take accurate bearings. You can do also find magnetic north using the North Star, the watch method or the Southern Cross (see below).

USING MAPS

The best maps to use are the British Ordnance Survey type. They are available in a variety of scales, such as 1:50000 and 1:25000 (remember to choose the one that fits your requirements). The most important skill is the ability to translate the lines on a map into the actual shape of the terrain.

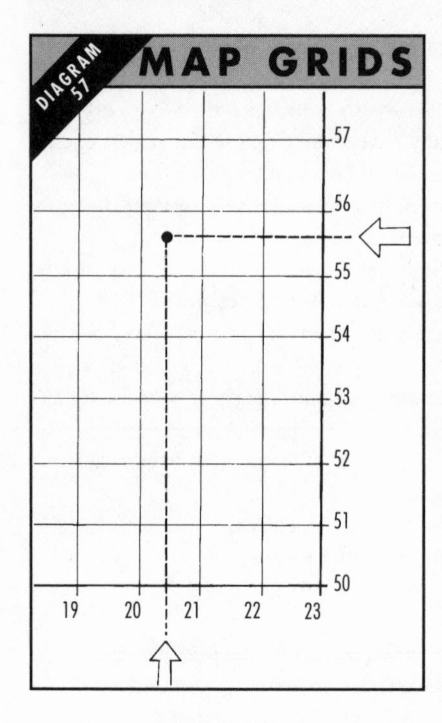

Grids on maps usually form squares to help you assess distances. You can find a position on a map by quoting grid coordinates. The reference is usually given as a six digit number (Diagram 57). To give readings always bear in mind that grid references adhere to the rule 'along the corridor, up the stairs.' The first three numbers are taken from the bottom or top margin, the second three from the left or right margin (you must mentally divide each map square into tenths to pinpoint the location). The map reference for the spot on the map in Diagram 57 is therefore 205556.

Symbols A knowledge of map symbols, combined with grids, scale and distance, give enough information for you to locate two points on a map and work out how long it will take you to travel between them.

Contour lines The undulations and relief of the land are represented on a map by imaginary slices at vertical intervals. By studying these contour lines you can build up a mental picture of the convex, concave surfaces and depressions of the land.

Contour lines indicate a vertical distance above or below a datum plane. Starting at sea level, each contour line represents an elevation above sea level. The contour interval is the vertical distance between adjacent contour lines (the distance of the contour level will be given in the map margin).

Contour lines are usually printed in brown, with every fifth contour being drawn in a heavier line. These heavier lines are called index contours, and somewhere along each one the line will be broken and its elevation given. The contour lines between index contours are called intermediate contours.

Using the contour lines on a map, you can find the elevation of any point by:

- ☐ Finding the contour interval of the map from the marginal information and noting the amount and unit of measurement.
- ☐ Finding the numbered contour line nearest the point for which elevation is being sought.
- ☐ Counting the number of contour lines that must be crossed to go from the

numbered line to the desired point and noting the direction (the distance above or below the starting value is calculated by multiplying the number of lines crossed by the contour interval.

Learn to identify the following terrain features on a map by the shape of their contour lines:

HILL: a point or small area of high ground. A hill usually slopes down on all sides, and its contours lines are like the ripples on a pond.

VALLEY: a stream course that has at least a certain amount of level ground, and which is bordered on all sides by higher ground. Contours indicating a valley can be U-shaped and run parallel to a major stream before crossing it.

DRAINAGE: a less-developed stream course in which there is no level ground and thus room for movement within its confines. The ground slopes upwards on each side and towards the head of the drainage. Contours indicating a drainage are V-shaped, with the point of the 'V' towards the head of the drainage.

DEPRESSION: resembles a hill but the contour lines are decreasing in height towards the centre of the feature.

The spacing of contour lines (Diagram 58) indicates the nature of the slope. Evenly spaced and wide apart lines indicate a gentle, uniform slope (A),

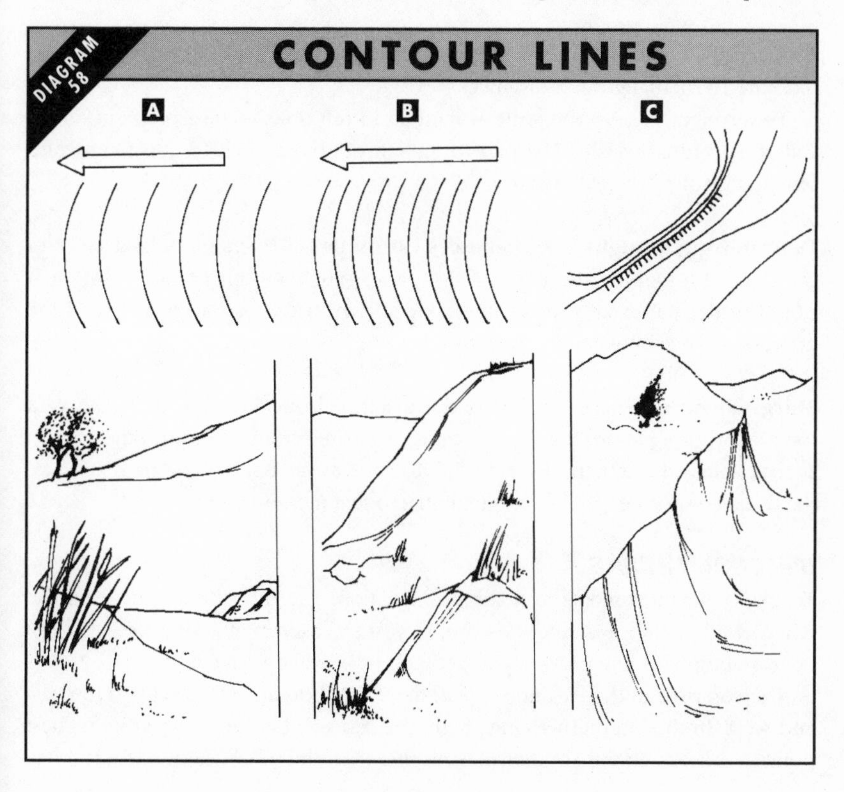

BRITISH SAS TIPS

MAKING YOUR OWN MAPS

If you do not have a map, use SAS training and plot your own. All you need is paper, something to write with and a keen eye.

- Find the best vantage point and examine the terrain.
- Note direction of ridges and count the number you can see.
- Make a general map with blank patches and then fill them in as you gain more information from other vantage points.
- Mark anything of note on the map, such as isolated trees and oddly shaped features.
- Use your map to mark your traps and areas where food and fuel can be found. It will be a great benefit to you and fellow survivors.

whereas lines evenly spaced and close together indicate a uniform steep slope (B). A vertical or near vertical slope is often shown by a ticked contour (C), the tick always point towards the lower ground.

Bar scales These are rulers printed on maps on which distances may be measured as actual ground distances.

To the right of zero the scale is marked in full units of measurement and is called the primary scale. The part to the left of zero is divided into tenths of a unit and is called the extension scale.

Determining straight-line distances on maps Lay a straight-edged piece of paper on the map so the edge of the paper touches both points. Mark each point on the paper and move the paper down to the bar scale and read the ground distance between the two points.

Marginal notes These often give the distance from the edge of a map to a town, road or junction. If a road distance is desired from a point on the map to such a point off the map, measure the distance to the edge of the map and add the distance specified in the marginal note to that measurement.

DIRECTION FINDING

To plan your movements in a survival situation, you must be able to establish where north, south, east and west are, so you can estimate the direction you will be travelling. This will prevent you getting lost or walking in circles.

The sun rises in the east and sets in the west, though *NOT* exactly in the east and west. In the Northern Hemisphere, the sun will be due south at its highest point in the sky. In the Southern Hemisphere, on the other hand, this noon day point will mark due north. The way that shadows move will indicate the hemisphere: clockwise in the north, anti-clockwise in the south.

In a survival situation, you can use some simple methods of determining both time and direction (though they all require the sun): by shadow (Diagram 59); by the equal shadow method (Diagram 60) and by a watch (Diagram 61). However, if you are using a watch you must be wearing one that has minute and hour hands, ie not a digital watch. The shadow-tip method is good for spot checks on your journey (it works at any time during the day when there is sunshine).

Shadow-tip method (Diagram 59) This method can also be used to determine the time. Move the stick to the intersection of the east-west line and the north-south line. Place it vertically in the ground. The west part of the east-west line show 0600 hours and the east part is 1800 hours, regardless of where you are on the planet.

The north-south line now becomes the noon line. The shadow of the stick becomes the hour hand in the shadow clock. The shadow may move either

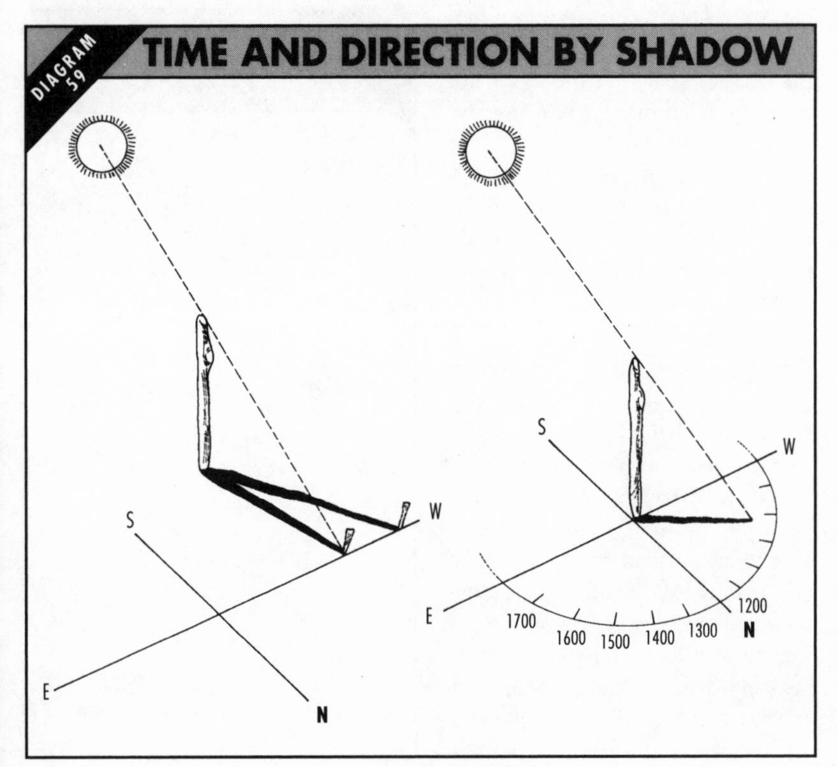

USMC TIPS

DETERMINING DIRECTION BY SHADOW (DIAGRAM 59)

The United States Marine Corps has a tried and tested method for determining location using just a stick and the shadow of the sun.

- Place a stick or branch in the ground at a level spot. Mark the shadow tip with a stone.
- Wait 10-20 minutes until shadow tip moves a few centimetres. Mark the new position of the shadow tip with a stone.
- Draw a straight line through the two marks to obtain an approximate east-west line (the sun rises in the east and sets in the west – the shadow tip moves in the opposite direction).
- Draw a line at right angles to the east-west line to get an approximate north-south line.
- Inclining the stick does not impair the accuracy of the shadow tip method, thus you can use it on sloping ground.

clockwise or anti-clockwise, depending on your location and the the time of year, but this does not alter your manner of reading the shadow clock.

The clock always reads 0600 hours at sunrise and 1800 hours at sunset. Nevertheless, it is a satisfactory way of telling time in the absence of a watch.

Equal shadow method (Diagram 60) is more accurate than the shadow-tip method (of which it is a variant) and consequently takes longer to carry out. Place a stick vertically into the ground at a level spot to cast a shadow at least 30cm (12in) long. Mark the first shadow tip in the morning (it is best to mark a shadow at least 10 minutes before the sun reaches its highest point).

Draw a clean arc at exactly this distance from the stick, using the

DIAGRAM 60

EQUAL SHADOW

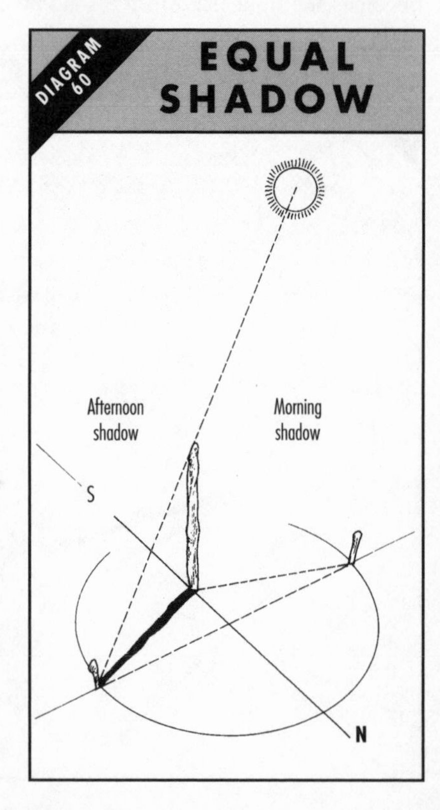

US ARMY TIPS

DETERMINING DIRECTION WITH A WATCH (DIAGRAM 61)

If you have a non-digital watch, wherever you are you can use it to determine north and south using the following military method.

- In the Northern Hemisphere, point hour hand towards the sun.
- A south line can be found midway between the hour hand and 1200 hours (if in doubt as to which end of the line is north always remember that the sun is in the east before noon and in the west in the afternoon).
- In the Southern Hemisphere, point the 1200 hours dial towards the sun. Exactly halfway between the 1200 hours dial and the hour hand will be a north line.
- Be sure to set you watch to true local time.

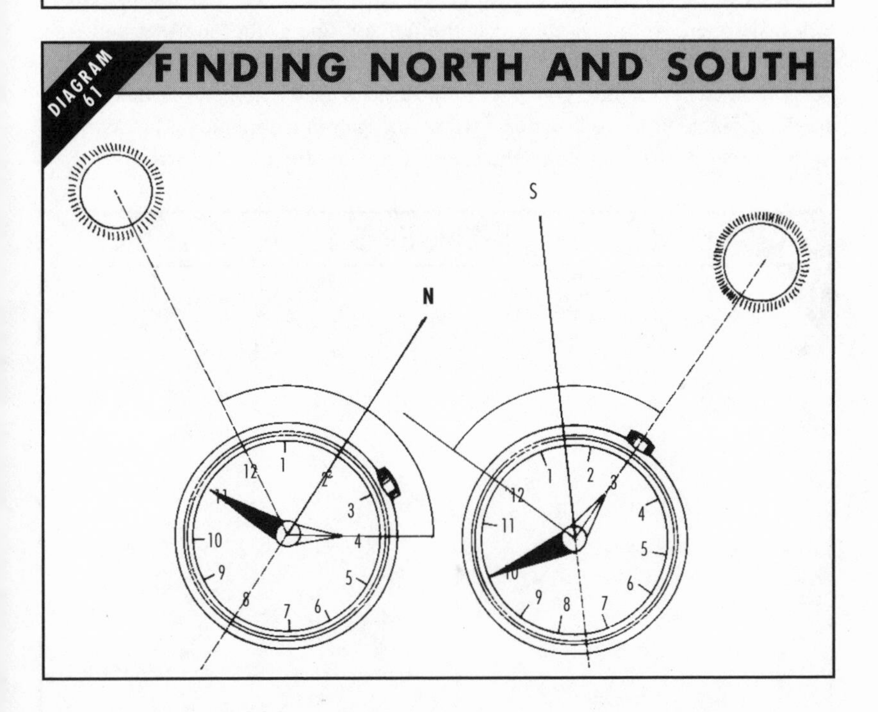

stick as the centre point. The shadow will shrink and move as midday approaches. After noon the shadow lengthens and will cross the arc. As it does so, mark the *EXACT* spot where it touches the arc. Then draw a straight line through the two marks to obtain an east-west line (west is the morning mark). Remember, you must mark the spot at the time the shadow touches the arc.

THE STARS

You can also use star constellations to determine direction. All survivors should know the following information concerning the bearings of stars:

POLE STAR (Diagram 62): in the Northern Hemisphere it is never more than one degree from the North Celestial Poles (NCP). This probably one of the most useful stars for determining direction.

BIG DIPPER: is very close to the NCP. Its two outer stars point directly to the Pole Star.

CASSIOPEIA: is also very close to the NCP.

SOUTHERN CROSS: an imaginary line through its long axis points towards the South Pole.

FALSE CROSS: a large cross of stars which lies near the Southern Cross.

TRUE CROSS: another name for the Southern Cross.

COALSACK: the dark region in the sky directly above the South Pole.

CELESTIAL EQUATOR: projection of the earth's equator into the imaginary celestial sphere. It always intersects the horizon line at the due east and west point of the compass. Thus, any star on the celestial equator rises due east and sets due west. Therefore, you can use these stars to determine your direction of travel at night, or to verify your daytime navigation techniques. The stars that you can see relatively easily on the celestial equator are listed below.

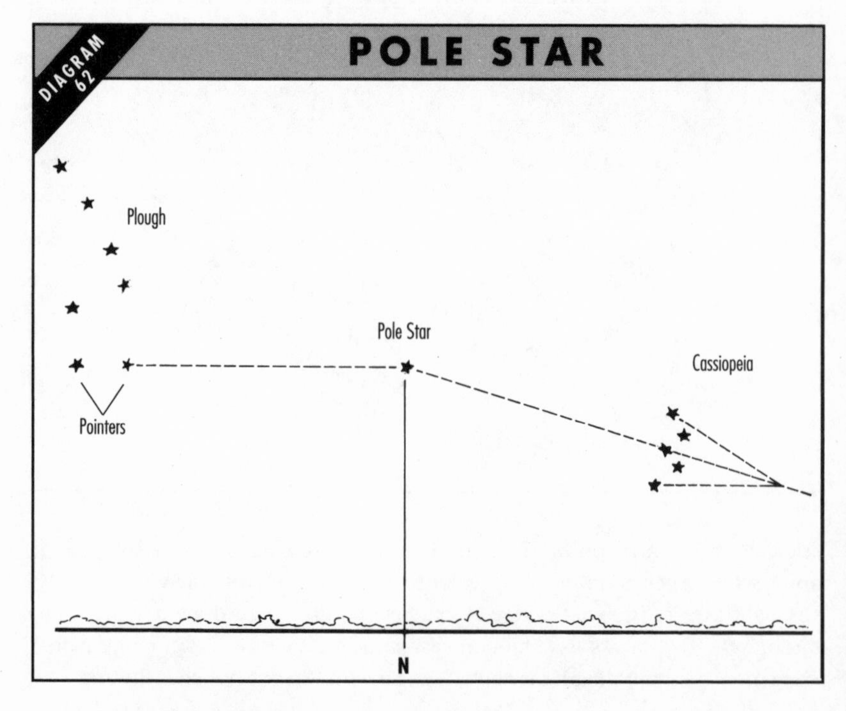

The celestial equator The following list gives the star groups which lie near the celestial equator and which are visible to the naked eye, together with the month of the year at which each particular star group is on the highest point of its path above the horizon.
JANUARY: Eridanus and Taurus.
FEBRUARY: Lepus, Orion and Monoceros.
MARCH: Canis Minor, Gemini and Cancer.
APRIL: Sextans and Leo.
MAY: Cater, Virgo and Corvus.
JUNE: Bootes.
JULY: Serpens (Caput), Libra and Ophiuchus.
AUGUST: Serpens (Cauda), Hercules and Scutum.
SEPTEMBER: Aquila and Delphinus.
OCTOBER: Capricornus, Equuleus, Pegasus and Aquarius.
NOVEMBER: Pisces.
DECEMBER: Cetus and Aries.

Orion (Diagram 63) rises above the Equator and can be seen in both hemispheres. It rises on its side, due east, regardless of the observer's position, and sets due west.

The Southern Cross (Diagram 64) In the Southern Hemisphere the Southern Cross, a constellation of five stars, can be used to determine south, though unfortunately it is not as easy to find as the Pole Star. To find south, project an imaginary line along the cross and four and a half times longer and then drop it vertically down to the horizon.

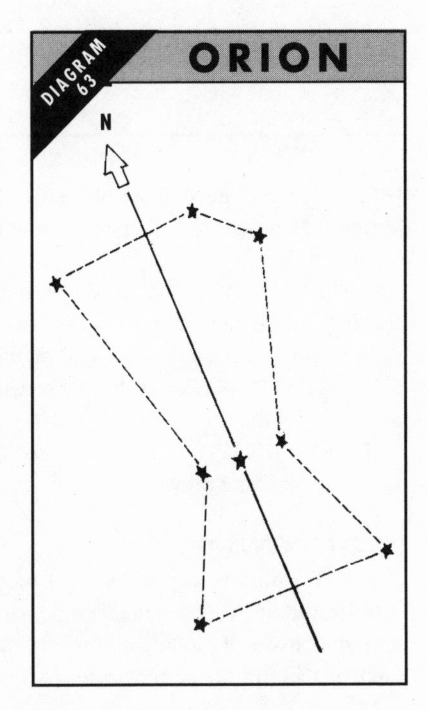

Star movement can be used to determine your position. If a star is observed over two fixed points for 15 minutes, it will be seen to move. In the Northern Hemisphere the following rules apply:
- ☐ If the star is rising, you will be facing due east.
- ☐ If the star is falling, you will be facing due west.
- ☐ If the star is looping to the right, you will be facing south.

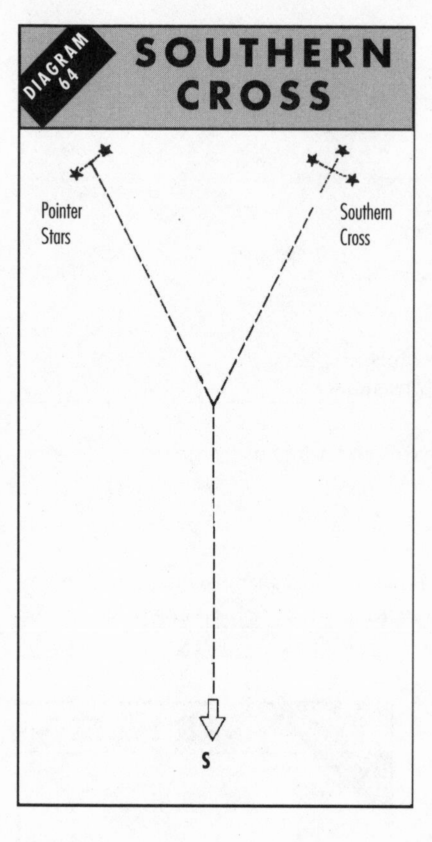

☐ If the star is looping to the left, you will be facing north.

Reverse these rules if you are in the Southern Hemisphere.

The moon If the moon rises before the sun sets, the illuminated side will be on the west. However, if it rises after the sun sets, the illuminated side will be on the east. If the sun rises at the same time as the sun sets it will be full and its position in the sky is east at 1800 hours, southeast at 2100 hours, south at 2359 hours, southwest at 0300 hours and west at 0600 hours.

NATURAL SIGNPOSTS

If you cannot see the sun, stars or the moon due to weather conditions, you can still determine direction by using natural signposts, though they are not as accurate and you should double check them and treat them with caution. Nevertheless, the following *GENERAL* rules apply:

TREES: normally grow most of their foliage on their sunny side, which in the Northern Hemisphere is the southern side and in the Southern Hemisphere the northern side.

CONIFERS AND WILLOWS: usually lean towards their sunny side.

FELLED TREES: their rings are widest on the northern side.

MOSS: tends to favour the dark and damp side of its host.

TREES WITH A GRAINY BARK: these usually have a tighter grain on the north-facing side of the trunk.

BIRDS AND INSECTS: build their nests in the lee of any cover (you must know the direction of the prevailing wind).

DEAD RECKONING

Dead reckoning is a good way of navigating a route from one location to another, though you will need some sort of writing implement and paper. The method consists of plotting and recording a series of courses before you set out, each one being measured in terms of distance and direction between two points. These courses lead from the starting point to your ultimate destination, and enable you to determine your position at any time, either by following your plan or by comparing your actual position on the ground in relation to your plotted course.

To navigate by dead reckoning you will need a number of aids: a map to select your route and for plotting your actual route as you are walking, a compass for direction-finding, a protractor for plotting direction and distance on the map, and a route card and log. A route card is used to outline the plan of your proposed journey and the log is used to record the distance you have actually travelled.

Having determined your starting point and plotted your route on the map, make out your route card. This describes each leg of the proposed route in terms of distance and direction. When you have completed your route card, you are ready to move. When you are walking you must keep a careful record of each bearing taken and the distance covered on each bearing. This record is your log: your memory is not enough. If you have to deviate from your route because of terrain, then you must make adjustments to your route and record them in the log.

When using dead reckoning, it is important to establish the length of your average pace. However, when you are computing your average pace remember to take into account the following:

SLOPES: the pace lengthens on a downward slope and shortens on an upward slope, though not too much on very gentle slopes.

WINDS: a headwind shortens the pace and a tailwind increases it.

SURFACES: gravel, mud, sand, long grass, deep snow and similar surface materials tend to shorten the pace.

ELEMENTS: snow, rain and ice all reduce the length of the pace.

CLOTHING: carrying excess weight of clothing shortens the pace and your shoe type can affect traction and thus your pace length.

SIGNALLING

As a survivor, it is important that you are able to give signals that a rescue team, specifically an aircraft, will be able to see clearly. You must plan your signalling system early on so you know how to use it and can do so at short notice. Unless you have a radio or flares, smoke and fire are your best ways to alert a rescue aircraft. Three fires or three columns of smoke are internationally recognised distress signals.

Commercially manufactured signals If you have come down in an aircraft or are in a life raft, you may have access to one or more commercially made signalling systems. If you are a backpacker, you are strongly advised to equip yourself with some of the items listed below:

TRANSCEIVER: can transmit tone or voice and will receive tone or voice.
BEACON: can only transmit tone.
RADIO: survival radios are generally line-of-sight communication devices, thus the best transmission range will be obtained when you are operating from clear, unobstructed terrain.
HAND-HELD FLARES: day flares produce a bright-coloured smoke; night flares are very bright and can be seen over long distances.
HAND-HELD LAUNCHED FLARES: are designed to overcome the problems of terrain masking and climatic conditions.
TRACER AMMUNITION: if available, it can be used for signalling. When fired, the round appears as an orange-red flash. *DO NOT* direct it at a rescue aircraft.
SEA MARKER: a rapid dispersion powder that stains the sea green or orange.
PAULIN SIGNALS: rubberised nylon markers that are blue on one side and yellow on the other.
WHISTLE: useful for short-range signalling.
LIGHT SIGNALS: torches or strobe lights can be seen over great distances.
SIGNAL MIRROR: a mirror flash can be visible up to a range of 60km (100 miles) in ideal conditions.

Smoke In daylight smoke will be recognisable over long distances. Signal fires should be built, covered and maintained ready to be lit at a moment's notice. Try to create smoke that will contrast with the background terrain. If you put green leaves, moss or damp wood on a fire you will get white smoke; rubber or oil-soaked rags on a fire will produce black smoke.

To build a smoke generator, make a large log cabin fire (see Fire Chapter) on the ground. This provides excellent ventilation and supports the green boughs used for producing the smoke. Place smoke-producing materials over the fire and ignite the fire when an aircraft is in the immediate vicinity. If you are in snow or ice terrain, build the fire on a raised platform above the wet ground, otherwise it will burn through the snow..

Fire Very effective for signalling at night. Build a fire that gives out a lot of light (see Fire Chapter). A burning tree is a good way of attracting attention. Pitch-bearing trees can be set on fire when green. For other types of trees, place dry wood in the lower branches, and set it on fire. The flames will ignite the tree foliage. Remember to select a tree apart from other trees – you do not want to start a forest fire!

Reflector On a sunny day, mirrors, polished canteen cups, belt buckles or other objects will reflect the sun's rays. Always practice signalling before you need it. Mirror signals can be seen for 100km (62 miles) under normal conditions and over 160km (100 miles) in a desert environment.

BRITISH SAS TIPS

SIGNAL FIRES

Learn these British SAS tips about where to build signal fires. You must get it right first time – you may not get a second chance!

- Keep green boughs, oil or rubber close by to create smoke.
- Build earth wells around fires if surrounded by vegetation or trees.
- Build fires in clearings. Do not build among trees: the canopy will block out the signal.
- If by a river or lake, build rafts to place fires on and anchor or tether them in position.

Ground-to-air signals (Diagram 65) There are several factors you must take into account with regard to ground-to-air signals if they are to be effective and help you get rescued. Above all, you must try to visualise what your signal will look like when viewed by a pilot from the air.

WARNING

Beware – do not direct a mirror beam on the cockpit of an aircraft for more than a few seconds – it can blind the pilot.

SIZE: signals should be as large as possible: remember they have to be seen from the air.
RATIO: signals should be proper proportion, especially if you are laying out letters. Always think of how they will look from the air.
ANGULARITY: make all pattern signals with straight lines and square corners (there are no straight lines or square corners in nature).
CONTRAST: the signal should stand out against the background.
ON SNOW: any dye used around the signal will add contrast.
ON GRASS: burn the grass to make a pattern.
ORANGE MATERIAL: tends to blend in, not stand out, when placed on a green or brown background.
OUTLINES: outline a signal with green boughs, brush or rocks to produce shadows, or raise a panel on sticks to cast its own shadows.
LOCATION: your signal should be placed where it can be seen from all directions (a large, open area is best).
MEANING: a signal should tell rescue service something pertaining to your overall situation.
Diagram 65 shows the internationally recognised emergency signals. Learn them, or better still carry a piece of paper around with you that lists them. When laying them out make them as large as possible, at least 10m (40ft) long and 3m (10ft) wide. At night, dig or scrape a signal in the earth, snow or sand, pour in

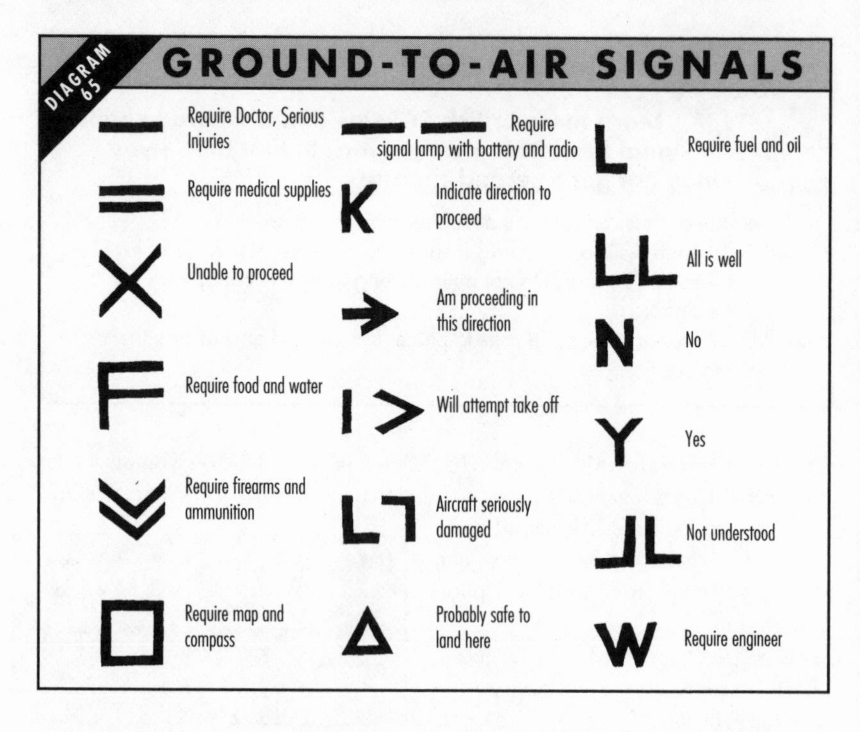

petrol and ignite it. This signal will be visible not only at night, but also during the day where the ground has been burnt.

You must destroy all ground-to-air symbols after rescue, otherwise they will go on marking after you have gone. Failure to do so may result in other aircraft spotting them and attempting a rescue.

Morse code (Diagram 66) You can transmit messages using Morse code by flashing lights on and off, using reflectors, by waving a flag a shirt tied to a stick (for a 'dot' swing to the right and make a figure-of eight, for a dash swing to the left and make a figure-of-eight), or using sound.

Body signals Diagram 67 shows a series of signals that will be understood by aircrew. It is advisable to use a cloth in the hand to emphasise the YES and NO signals. Note the changes from frontal to sideways positions and the use of the legs. Whenever you are making signals, always do so in a clear and exaggerated way (remember you will be a great distance from the aircraft).

An aircraft that has understood your message will tilt its wings up and down in daylight or make green flashes with its signal lights. If the pilot has not understood your message, he will circle his aircraft during daylight or make red flashes with his signals lights at night. Once a pilot has received and understood

DIAGRAM 66

MORSE CODE

A	•–	M	––	Y	–•––
B	–•••	N	–•	Z	––••
C	–•–•	O	–––	1	•––––
D	–••	P	•––•	2	••––
E	•	Q	––•–	3	•••––
F	••–•	R	•–•	4	•••–
G	––•	S	•••	5	•••••
H	••••	T	–	6	–••••
I	••	U	••–	7	––•••
J	•–––	V	•••–	8	–––••
K	–•–	W	•––	9	––––•
L	•–••	X	–••–	0	–––––

DIAGRAM 67

BODY SIGNALS

your first message, you can transmit other message. Be patient: don't confuse the person flying the aircraft.

Shadow signals can be very effective for signalling, though you must construct them in the proper way. Make sure you build them in a clearing, of a sufficient size, and that they contrast with their immediate surroundings. Follow these construction guidelines when making shadow signals in the following terrain :
ARCTIC WINTER: build a snow block wall, line the blocks alongside the trench from which the blocks were cut.
ARCTIC SUMMER: construct walls from stones, sods or wood.
WINTER BELOW THE SNOW LINE: stick green boughs in the snow and build a wall of brush and boughs around them.
SUMMER BELOW THE TREE LINE: use piles of rocks, dead wood, logs and sod blocks cut out from the earth.

Mountain rescue code The codes listed below are internationally recognised mountain rescue signals. Learn them and practise transmitting them (you should always carry something to enable you to transmit a signal).

US ARMY TIPS

SIGNALS USING NATURAL MATERIALS

When you are stranded in the wild, follow US Army advice and construct signals from the materials around you.

- Build brush or snow mounds that will cast shadows.
- In snow, tramp down the snow to form letters or symbols and fill in with contrasting materials: twigs or branches.
- In sand, use boulders, vegetation or seaweed to form a symbol.
- In brush-covered areas, cut out patterns in the vegetation.
- In tundra, dig trenches or turn the soil upside down.
- In any terrain use contrasting materials so that the symbols are visible to aircraft.

SOS To send this signal use the following flare, sound and light signals:
FLARE: red.
SOUND: three short blasts, three long blasts, three short blasts (repeat after a one-minute interval).
LIGHT: three short flashes, three long flashes, three short flashes (repeat after a one-minute interval).
Help needed To send this signal use the following flare, sound and light signals:

FLARE: red.
SOUND: six blasts in quick succession (repeat after a one-minute interval).
LIGHT: six flashes in quick succession (repeat after a one-minute interval).

Message understood To send this signal use the following flare, sound and light signals:
FLARE: white.
SOUND: three blasts in quick succession (repeat after a one-minute interval).
LIGHT: three flashes in quick succession (repeat after a one-minute interval).

Return to base To send this signal use the following flare, sound and light signals:
FLARE: green.
SOUND: prolonged succession of blasts.
LIGHT: prolonged succession of flashes.

Information signals (Diagram 68) are used when you leave the scene of a crash or abandon camp. Always leave a large arrow to indicate the direction in which you have set off. All your trails should be marked

RESCUE

Once the rescue services have established your position, a rescue operation will be launched. On land, this will usually be in the form of an aircraft or helicopter, and a ship or boat if you are in the water. Try to give consideration to the recovery site if you can: be aware of the effects of weather and terrain on the rescue aircraft. Try to avoid overhangs, cliffs or the sides of steep slopes if possible.

Helicopter rescue Helicopters will make a rescue by landing or hovering. Landings are usually necessary on high ground, where a helicopter has insufficient power for hovering. Hoist recovery is the preferred method for effecting a water recovery. If a helicopter has to land to pick you up, do not approach it from the rear, it is a blind spot for the crew and the tail rotor is unprotected. Similarly, be careful of being hit by the rotors if you are approaching the helicopter down a slope. Be aware of any materials on the ground that may be sucked into the rotor blades, such as a parachute or tent. Pick them up before the helicopter lands, including any leaves and twigs.

Try to mark the landing site with a large 'H' (at least 3m (10ft(high), and however you make it remember to make sure the materials are securely anchored to the ground.

If you are in sandy terrain, try to water the sand to keep dust down. In snow, try to compact the surface down as much as possible (soft, wet snow will cling to the aircraft, and powdery snow will swirl and restrict the pilot's vision).

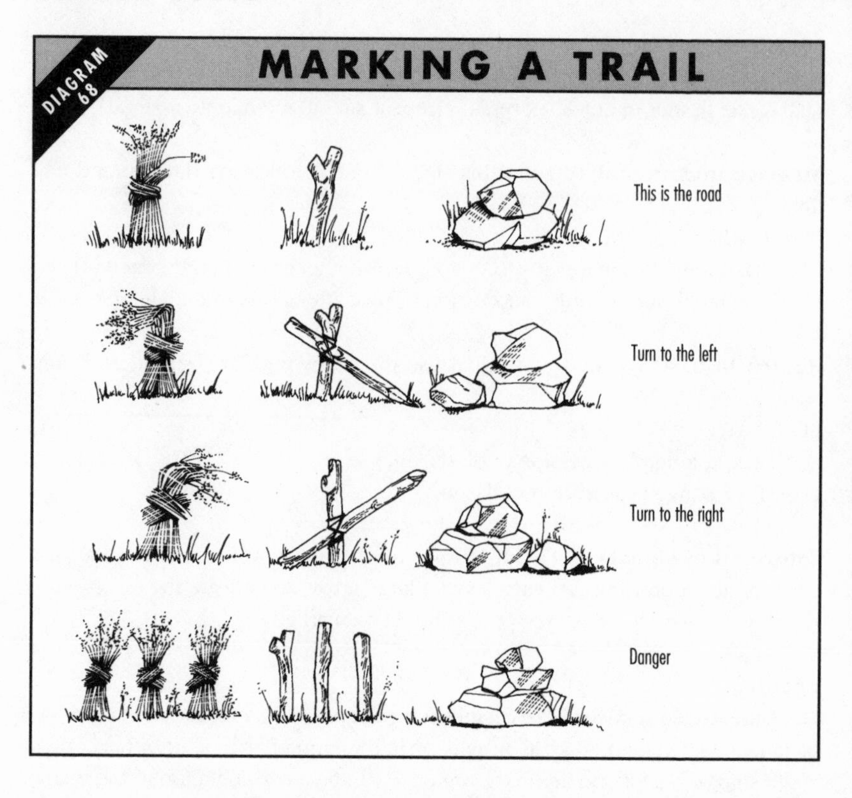

If you are being winched aboard a hovering helicopter, let the cable touch the ground before your touch it (all aircraft build up a lot of static electricity, and you will receive a shock if you grab it before it discharges through the earth). Fit yourself into the strop and then give the 'thumbs up' sign. Do not make any further signals, especially raising your arms – you could slip out of the strop.

When you get to the cabin door, let the winchman do everything and obey him to the letter. When you are out of the winch you will be told to sit down and shown where to sit. Go immediately to this place to be fastened into the seat. Remember to obey the aircrew at all times. Be careful when you leave the helicopter after it has touched down: wait for the rotors to stop turning – obey crew members at all times.

Aircraft rescue If an aircraft is able to land to pick you up, do not approach it until it has come to a halt and the pilot or another member of the crew has signalled you to do so.

Once you have been rescued, you should, where relevant, inform the rescue services at once of any survivors that have become separated from the group.

THE FORCES OF NATURE

Unless properly harnessed, nature is the survivor's most formidable enemy. Ice floes, mountains, deserts, jungles and oceans can all be killers, with each posing its own distinct set of threats. You must learn how to cope in every type of hostile environment.

POLAR REGIONS

Freezing temperatures, chilling winds and seemingly barren terrain – these are just some of the threats you face in cold-weather areas. Learn to use the resources at your fingertips to live in this unforgiving environment.

CHARACTERISTICS OF THE TERRAIN

There are two types of cold-weather areas: snow climates and ice climates. In addition to the incessant cold, the other great threat to survival is the wind. In the Antarctic, winds of up to 177km/hr (110mph) have been recorded. The combination of low temperatures and the wind creates a condition known as windchill. For example a 32km/hr (20mph) wind will bring a temperature of -14 degrees C (5 degrees F) down to -34 degrees C (-30 degrees F). This can pose great danger to the survivor: exposed flesh can freeze in seconds.

Snow climates are those areas that lie between 35 and 70 degrees north latitude. The tree line is the best boundary to define where snow climates end: to the north of it lies the tundra of the frozen wastes. Snow climates themselves are divided into two separate types: continental subarctic and humid continental.

The continental subarctic is characterised by vast extremes, with temperatures ranging from -43 degrees C (-45 degrees F) to 44 degrees C (110 degrees F). These areas are cold, snowy and moist all year, with cold, short summers. Winter is the dominant season of the continental subarctic climate. In such a climate, where freezing temperatures occur for 6-7 months, all moisture in the ground is frozen to a depth of several metres.

> **DANGER**
>
> ***Windchill can freeze your flesh and can be life threatening. Be aware that wind is created when you walk, run or ski, and can cause frostbite.***

The humid continental climates (the northern United States, Canada and central Asia) are located between 35 and 60 degrees north latitude. Being a battleground between polar and tropical air masses, they are subject to strong seasonal contrasts and variable weather. Summer temperatures range from 16-21 degrees C (60-70 degrees F) and winter from -9-4 degrees C (15-25 degrees F), with heavy snow fall.

Both climates have seasonal extremes of darkness and daylight. Generally speaking, the nights of snow climate areas are long, even continuous, in winter. This can be a problem if you are a survivor: no heat is received directly from the sun and so the temperatures are very cold. The lack of light also restricts the amount of activity you can undertake outside your shelter, though the light from the moon, stars, and auroras (streamers of light) reflecting off the white snowy ground does help. In addition, continued confinement in a cramped shelter can mean you start to get bored and depressed. In such a situation you must remember that the periods of complete darkness do not last long.

Snow areas are covered by needle-leaf forests, with an abundance of lakes and swamps. The coastlines vary from gentle plains sweeping down to the sea, to steep rugged, cliffs with glaciers at high altitudes. Vegetation ranges from cedar, spruce, fir and pine trees to dwarf willow, birch and alder nearer the tundra line.

Ice climates There are three types of ice climates: marine subarctic climate, tundra climate and ice cap climate.

Marine subarctic climates are characterised by cloudy skies and strong winds. The area lies between 50-60 degrees north latitude and 45-60 degrees south latitude. The tundra climate region lies north of 55 degrees latitude and south of 50 degrees latitude. The temperature never rises above -10 degrees C (14 degrees F) and there is cloud cover for most of the year. There are three vast areas of ice on the Earth, which comprise the ice cap regions: Greenland, the Antarctic and the Arctic.

The climate is harsh in the extreme: vast rugged mountains, steep terrain, snow and ice fields, glaciers and high winds. There is some wildlife, such as Arctic birch and shrubs, plus herbaceous plants: grasses, black crowberry and cowberry. In addition, there are mosses and lichens. All tundra plants are small in stature compared to those that grow in more southerly regions (though in valleys and along rivers they may reach the height of a man), and they tend to spread along the ground to form large mats.

MOVEMENT

Because of the harshness of the terrain, movement should only be undertaken if your present location is hazardous or if you are near civilisation and the possibility of rescue. If you are moving by foot you will only be able to take with you what you can carry. You will burn off a lot of calories and sweat more water than normal (both of which will have to be replaced). Improvised snow shoes (Diagram 69) are essential. Use willow or other springy wood (A), fashion cross-pieces (B) and toe and heel straps (C).

US ARMY TIPS

TRAVEL IN SNOW AND ICE AREAS

Military operations in snow and ice areas can be fraught with difficulties, not the least of which are the problems associated with the unpredictability of the weather. The US Army has drawn up the following guidelines:

- 'Whiteouts' (complete snow cover and clouds so thick and uniform that light reflected by the sun is about the same intensity as that from the sky) can occur. This can result in survivors falling into crevasses, over cliffs or high snow ridges.
- Poor visibility makes navigation difficult. A compass is a necessity, but because of magnetic variation navigating a true course is difficult.
- In summer, there will be a mass of bogs, swamps and standing water, which are all difficult to cross. They are accompanied by a mass of mosquitoes which can inflict severe discomfort if body parts are not covered (use insect repellent if you have it).
- In mountainous terrain there may be crevasses and glacial streams, which the survivor may fall into. Use a stick to probe ahead.
- In timbered areas, travel will be easier on skis or snow shoes during the cold months.

DIAGRAM 69

IMPROVISED SNOWSHOES

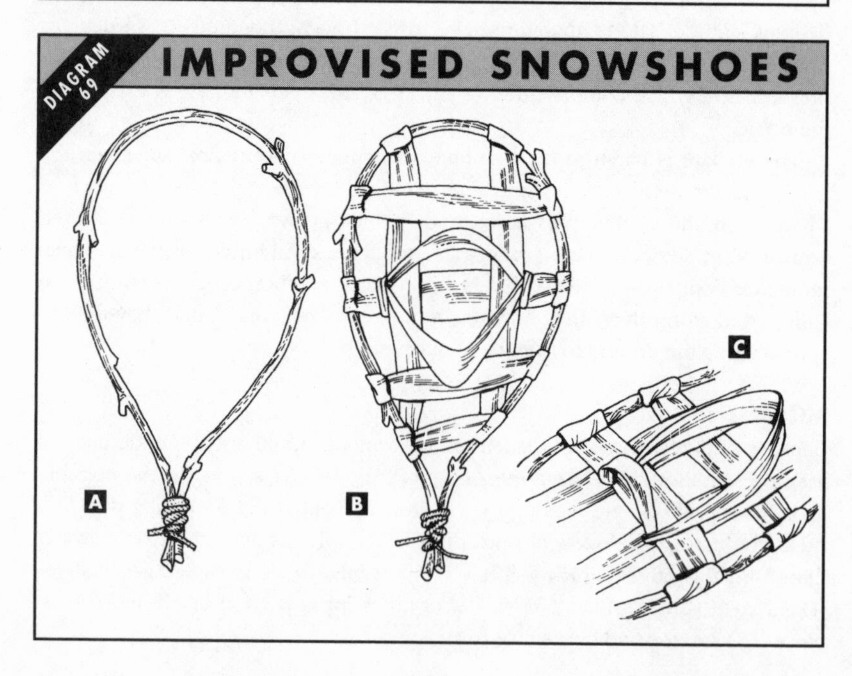

Leaving camp If you decide to leave camp you must leave some sort of permanent signal to indicate to rescue crews that you have left camp and the direction in which you headed. The snow block shadow signal (Diagram 70) is the best signal in snow conditions. It should be as large as possible in an open area and should point exactly in the direction you intend to travel. In addition, you should build further signals along your route so they can be followed.

Signalling Carry a reflector in a handy place in case an aircraft appears. Alternatively, carry fire-starting material in you pocket to start a signal fire. Always have flares ready if you have them. Remember, speed is of the essence. You may only have one chance: don't ruin it by not being prepared.

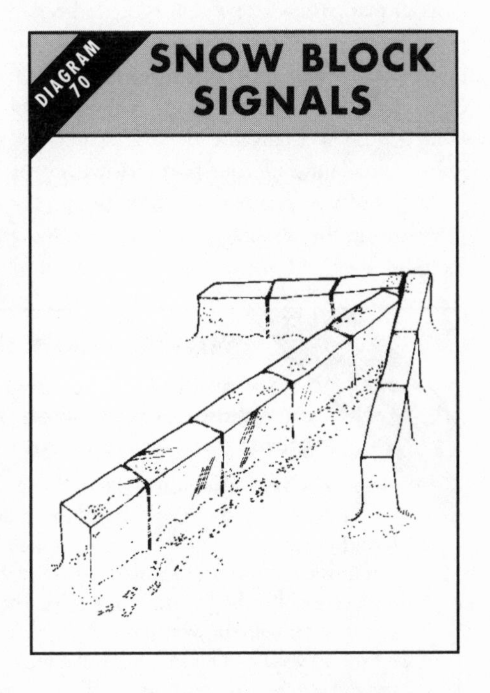

DIAGRAM 70 SNOW BLOCK SIGNALS

Navigation Unfortunately, compasses are unreliable near the Poles. Use the stars at night and the shadow-tip method during the day (see Signalling and Navigation Chapter).

Choose your route carefully at the start of each day. You do not want to get cold and wet by going through swamps and bogs. Try to follow a waterway if possible: most settlements are situated on a river or stream. If you follow a waterway you will be able to replace the fluids you lose through exertion. There will be fish in the river or stream, and animals will be attracted to it to drink, giving you the opportunity of catching them. In addition there will probably be an abundance of edible plants growing alongside it.

When following a waterway, resist the temptation to build a raft and float on it. Many northern rivers are fast, cold and dangerous and can smash a raft into splinters. You do not know the river, and even if one appears calm, remember that there could be rocks under the water that could tear your raft to pieces in seconds. You must bear the following points in mind when navigating in snow and ice areas:

- Poor or non-existent roads mean they are of little use for cross-country navigation.
- In winter, long nights, blowing snow and fog limit visibility.

- Snow falls can obliterate tracks and landmarks, thereby increasing navigational mistakes.
- There is limited daylight in the winter.
- Magnetic disturbances are common, making compass readings unreliable.
- Large-scale maps could be non-existent for the area you are in.
- You may encounter a multitude of lakes, ponds and creeks that are not on your map. This can be confusing and can lengthen travel times.

Travel clothing and equipment (see the section on cold-weather clothing in the What to Wear Chapter). You should protect the whole body from the cold and wind, especially the head and feet. By keeping active you will keep the blood circulating in your body. However, try not to sweat. If you start to over heat, loosen or remove some clothing. Try to keep clothing clean: dirt and grease clog the air spaces in your clothing and reduce insulation. Try to wear goggles, they will stop you getting snow blindness.

BRITISH SAS TIPS

NAVIGATION IN POLAR REGIONS

Follow SAS advice and learn to use only reliable navigational aids when travelling in polar regions: use nature to your advantage.

- When travelling on sea ice, do not use icebergs or distant sea landmarks to get your bearings. Floe positions can change.
- Avoid icebergs, which have most of their mass below the water. They can turn over without warning, especially with your weight on them.
- Keep clear of sailing close to ice cliffs. Thousands of tons of ice may fall into the sea without warning.
- Migrating wildfowl fly to land in the thaw, and most seabirds fly out to sea during the day and return at night, thereby indicating land.
- Clouds appear black underneath when over open water, timber or snow-free ground, white over the sea ice and snow fields. New ice produces greyish reflections.

Travelling on foot through snow and ice can be exhausting. Skiing is best on snow with a hard crust, but snow shoes are better for deep, lose snow. You can fashion tree branches to make a pair of snow shoes. When you walk with the shoes on, the binding should hinge at the toe so you can drag the tail end.

Finally, remember to protect your fingers. Keep you hands covered. Do not place them on metal when extremely cold, you will get a cold burn. If your hands get cold, place them inside your clothing under your armpits, next to your stomach, or in your crotch. Clenching and unclenching your hands in your mittens will also keep them warm.

CANADIAN AIR FORCE TIPS

POLAR TRAVEL HINTS

Canadian Air Force pilots operate over hostile terrain in the north of their country. If they have to bale out, they must know ice travel skills.

In summer:

- Avoid dense vegetation, rough terrain, insects, soft ground, swamps and lakes.
- Cross glacier-fed streams in the early morning to avoid raging torrents.
- Travel on ridges and game trails, maintaining constant direction checks.

In winter:

- Do not travel in a blizzard or during extremely cold weather. Camp and save your strength.
- Be wary of thin ice, heavy snow and air pockets if travelling on frozen rivers. Always use a pole to test the ground ahead.

DANGERS

The main threat to your survival is the intense cold. You must get out of the cold and the wind. Build a shelter (see below). You must also take care of your health, otherwise you could find yourself falling victim to a wide range of ailments, including dehydration, frostbite, sunburn and trench foot.

Dehydration The first indication that you are dehydrated is the colour of your urine: it will appear dark yellow. Other symptoms are no appetite, slow motion, drowsiness and a higher than normal temperature.

You must drink plenty of liquids – water, tea, soup – in cold weather to avoid dehydration. Avoid alcohol: it can dull your mind and lead to frostbite.

Hypothermia is the cooling of the body below its normal temperature of 36-38 degrees C (97-100 degrees F), and is potentially life-threatening. The symptoms are (in order): mild shivering; uncontrollable shivering (you will have difficulty controlling your fingers and hands); violent shivering (difficulty speaking); shivering slows down and stops; irrationality; inability to make decisions; unconsciousness and then death.

Keep watch for any signs of shivering. Be particularly alert for when the shivering slows down or stops. This is a critical warning.

The treatment for hypothermia is heat. Stop what you are doing get out of the cold. Build a fire, get dry, drink warm fluids, eat chocolate. *DO NOT* drink alcohol: it will only result in the loss of core heat.

If you come across a person who is hypothermic, get him or her out of the wind and into shelter. Remove wet clothing and replace with dry items. Place

US SPECIAL FORCES TIPS

HYGIENE IN POLAR REGIONS

You must keep clean in snow and ice regions. Follow US Green Beret guidelines and stay healthy and hygienic.

- Change socks and wash feet daily. If not possible remove boots and socks and massage and dry your feet (sprinkle with foot powder and then rub off).
- Clean teeth daily. If toothbrush is unavailable use a clean piece of cloth wrapped around a finger or end of a twig chewed into a pulp.
- Change underwear at least twice a week. If not possible to wash it, crumple it, shake it and air it for at least two hours.

warm rocks or water bottles filled with warm water near patient's throat, armpits and groin. Build a fire to provide heat. You may have to perform CPR (see First Aid Chapter).

Trench foot An injury caused by exposure to the cold and wet. In the early stages, feet and toes appear pale and feel numb, cold and stiff. Walking is difficult and the feet swell and become painful. You must be alert to prevent development of trench foot: it can lead to amputation. To prevent trench foot, make sure you clean and dry your socks and boots regularly, and dry your feet as quickly as possible if they get wet. If wearing wet boots and socks, exercise your feet continually by wriggling your toes and bending your ankles. Get out of them as quickly as possible.

When treating trench foot handle the feet very gently. *DO NOT* rub or massage them. Clean carefully with soap and water, dry and then elevate. Do not walk if you have trench foot.

Frostbite Frostbite is broken down into three categories of seriousness: frostnip, superficial frostbite and deep frostbite. It can be very serious, leading to the loss of toes, fingers, arms, legs and life. Factors that lead to frostbite include inadequate clothing, fatigue, too much alcohol, restricted circulation, long exposure and low temperatures.

Frostnip usually happens to the tips of ears, fingers, toes and the nose. The skin turns white but there is no pain.

Superficial frostbite strikes fingers, hands, toes, feet and face, sometimes knees and upper legs. The skin appears white, waxy and firm. The area will be numb, and may have a blue or purple outline.

Deep frostbite is a complete freezing of a part of the body. The affected area will be white and hard and completely numb.

BRITISH SAS TIPS

PREVENTING FROSTBITE

SAS soldiers often go on exercises in the arctic. They know that frostbite is a dangerous enemy, and they take every precaution to prevent it.

- Wrinkle your face to stop stiff patches forming and exercise hands.
- Keep an eye on yourself and others for patches of white, waxy skin, especially on the face, ears and hands.
- Do not wear tight clothing: it will restrict circulation.
- Dress inside your sleeping bag.
- Always wear the proper clothing outdoors. If it gets wet dry it out as quickly as possible.
- Brush snow off clothing before entering shelter, otherwise it will melt and wet clothing.
- Keep hands dry. Wear gloves and don't touch metal with bare hands: if you do you will suffer cold burns.
- Take care if you are very tired and rest if you are sick.

To treat frostnip: cover the affected area. To treat other frostbite: *NEVER* thaw an area if there is a chance of refreezing. When you are in a safe location you can thaw the affected parts. Gently soak in water that is kept at a temperature of 100-110 degrees F (it feels warm to the touch). The flesh should turn to pink or red: at this point you will experience extreme pain. Large blisters will form in a day: *DO NOT* lance them. They will break in 2-3 weeks. When they do break treat the area like a burn (see First Aid Chapter).

With deep frostbite, a hard black shell will form over the area. Leave it: it is protecting damaged tissue and will come off on its own in 3-6 weeks. The area should heal totally within six months to a year. Remember: do not thaw slowly in cold water; do not thaw by holding close to a fire; do not rub the area, especially with snow.

Immersion in water If you fall into water you must take immediate action, otherwise you could be dead within minutes (it will literally knock the breath out of you, and you will suffer loss of muscular control and violent shivering). As soon as you fall into the water, move as quickly and violently as you can for land. Once on land, roll in the snow to absorb water, get to shelter and into dry clothes. *SPEED IS OF THE ESSENCE.*

Sunburn You can get this very easily because the rays of the sun reflect upwards from snow and ice. Vulnerable areas are lips, eyelids and nose. Apply sunburn cream to these areas (including the inside of your nose).

Snow blindness Wear sunglasses. If you do go snow blind, your eyes will be red and sore; they will water and you will have a headache. Treatment includes blindfolding yourself and waiting until the soreness disappears. If you don't have sunglasses, improvise a pair from cardboard or tree bark (see Improvising Tools and Weapons Chapter).

Animal dangers Bears are present in the northern forests and wastes. Avoid them if you can: they are powerful and dangerous and can kill. Polar bears are found in the high Arctic. They are powerful and tireless. Avoid them unless you have a gun. Other animals to be avoided are walruses and elephant seals.

Plant dangers Do not eat water hemlock, baneberry fruit, arctic buttercups, lupin, larkspur, locoweed, false hellebore or death camas. They are all very poisonous (see Food Chapter).

FINDING WATER AND FOOD

The survivor needs a constant supply of food and water to sustain himself in polar regions. In particular, the construction of shelters and signals – even simple movement – results in increased physical exertion, which increases food and fluid requirements.

DIAGRAM 71 WATER MACHINE

WATER

Fortunately, there is an abundance of water in polar regions: streams, lakes, ponds, snow and ice. Remember to purify all surface water. If you let the water stand for a while any silt or dirt will settle on the bottom. Do not eat unmelted snow or ice, it lowers the body's temperature, induces dehydration and causes mild cold injury to legs and mouth membranes.

Construct a water machine (Diagram 71) to turn snow and ice into drinking water. Place snow on any porous material, gather up the edges and suspend it from a support over a

US ARMY TIPS

ICEBERGS AS A WATER SOURCE

Icebergs, because they are composed of freshwater, offer the survivor a potential source of drinking water in polar regions, but beware. Follow these US Army guidelines:

- n Exercise caution: even large icebergs can roll over and dump survivors into the sea (most of their mass is below the water, which can be potentially fatal.
- n Do not drink water obtained from fresh sea ice. Use old sea ice which is bluish or blackish and shatters easily – this is salt-free.
- n Snow and ice may be saturated with salt from blowing spray. If it tastes salty, discard.

container near a fire. The heat will melt the snow and the water will drip from the bag into the container.

FOOD

In snow and ice areas there are many types of food, both plant and animal, available to the survivor. However, you need to know where they can be located and when they are available.

Plants The following arctic and northern plants should be a part of your diet, especially lichens, which have sustained many survivors in cold climates.
RED SPRUCE Appearance: reaches up to 23m (70ft) in height and has yellow-green needles around its hairy twigs.
Edible parts: young shoots can be eaten raw or cooked. Eat the inner bark after boiling and infuse the needles to make tea (see First Aid Chapter).
BLACK SPRUCE Appearance: smaller than the red spruce with shorter needles.
Edible parts: young shoots can be eaten raw or cooked. Eat the inner bark after boiling and infuse the needles to make tea.
LABRADOR TEA Appearance: an evergreen shrub which has an average height of 30-90cm (1-3ft). It has narrow leaves with rolled edges, whitish or grey underneath, and five-petalled white flowers.
Edible parts: the leaves make a refreshing tea.
ARCTIC WILLOW Appearance: round leaves, shiny on top, and yellow catkins.
Edible parts: the spring shoots, leaves, inner bark and peeled roots are all edible.
FERNS Appearance: green.
Edible part: eat only the young fiddleheads up to 15cm (6in) long. Steaming is the best preparation.

SALMONBERRY Appearance: looks like a small wild raspberry. Has thornless, three-part leaves, purplish-red flowers and juicy red or yellow flowers.
Edible parts: the berries can be eaten raw.
BEARBERRY Appearance: mat-forming, woody, with club-shaped evergreen leathery leaves, pink or white flowers and clusters of red berries.
Edible parts: the plant is edible when cooked.
ICELAND MOSS Appearance: leathery, with grey-green or brownish mats up to 10cm (4in) high, composed of many strap-shaped branches.
Edible parts: soak all parts for several hours and then boil thoroughly.
REINDEER MOSS Appearance: a lichen that grows 5-10cm (2-4in) high in large clumps with hollow, roundish, greyish stems and branches that resemble reindeer antlers.
Edible parts: soak all parts for several hours and then boil thoroughly.
ROCK TRIPES Appearance: lichens that form roundish, blister-like greyish or brownish growths attached to rocks.
Edible parts: soak all parts for several hours then boil thoroughly.

Animals There are many animals that can act as a food source in snow and ice areas, though remember that some of the larger ones can be dangerous and should be avoided unless you have a firearm (see above).
CARIBOU: migrate throughout northern Canada and Alaska. In northern Siberia they migrate inland to nearly 50 degrees north latitude. Some are found in west Greenland. In summer they move close to the sea or into the high mountains, and in winter they feed on the tundra. *BE CAREFUL OF THEIR ANTLERS.*
REINDEER: found throughout northern Canada, Alaska and parts of Siberia and Greenland. Like the caribou, they spend summer close to the sea or in the high mountains and fed off the tundra in winter. *TAKE CARE: REINDEER ANTLERS CAN GOUGE AND STAB WITH GREAT POWER.*
MUSK OX: musk oxen are found in northern Greenland and on the islands of the Canadian archipelago. *BE WARY OF THESE ANIMALS, ESPECIALLY OLD MALES: THEY CAN BE DANGEROUS.*
SHEEP: can be found in snow regions. In the winter they descend to the valleys and lower areas.
WOLVES: they are present in snow regions and usually roam in pairs or packs.
FOXES: present in snow regions. They are seen most frequently when lemmings and mice are abundant.
BEARS: hibernate during the winter. Avoid these powerful creatures.
ARCTIC HARE: found in tundra regions. Can be shot or trapped in both winter and summer.
LEMMINGS: also inhabit tundra regions. They can easily be caught in traps.
SQUIRREL: live in tundra regions and prefer cover. Set your snares in shallow ravines. Squirrels hibernate in the winter. In summer ground squirrels are abun-

CANADIAN SPECIAL FORCES TIPS

HUNTING IN COLD WEATHER AREAS

Canada's winter warfare specialists are experts at living off the land. Learn the tricks they use when hunting animals.

- You can get near caribou by crawling on all fours. This may also attract a wolf, which will think you are a four-legged animal. Kill it.
- Moose can be found in heavy brush or crowding the lake shorelines.
- Mountain goats and sheep can be approached by quietly moving downwind while they are feeding with their heads lowered.
- Musk-ox leave cattle like tracks and droppings.
- Rabbits often run in circles when chased and return to the same place.
- Approach seals when they are sleeping. Keep downwind.
- Grouse and ptarmigans are 'tame' and can be approached easily.

dant along the sandy banks of large streams. *SQUIRRELS ARE SAVAGE WHEN DEFENDING THEMSELVES. THEIR TEETH CAN INFLICT A NASTY BITE.*

MARMOTS: live in the mountains among the rocks. Look out for a large patch of orange-coloured lichen on rocks – this indicates their burrows.

DUCKS: build their nests in the summer near ponds on the coastal plains or near lakes and rivers.

SWANS: nest on small, grassy islands during the summer.

GEESE: crowd together near large river or lakes in the summer.

GROUSE: common in the swampy forest regions of Siberia; they are also present in the winter.

PTARMIGANS: found in the swampy forests of Siberia. They are often seen in pairs or flocks, feeding along grassy or willow-covered slopes.

SEALS: they are vulnerable when they are on the ice floes and when they have their young (the pups are produced between March and April). Newborn pups cannot swim and are easy to catch.

CRANES: nest in and around the swamps, bogs and lakes of the tundra.

OWLS: are present in ice areas all year round.

WALRUSES: abound on the ice. *THOUGH THEY APPEAR CUMBERSOME, THEY CAN BE EXTREMELY AGGRESSIVE AND DANGEROUS. LEAVE THEM ALONE UNLESS YOU HAVE A GUN.*

Try to preserve some of your meat for future use if you kill a large animal or catch a lot of small ones. Freezing is the best way of preserving fresh meat or fish. Remember to suspend it off the ground, beyond the reach of scavengers. Always cook meat thoroughly and *DO NOT* eat the livers of seals or Polar bears – they contain dangerous concentrations of Vitamin A. Remember to bleed, gut and skin any carcass while it is still warm. Do not eat a diet that consists solely of rabbit or hare – it can kill you (see Food Chapter).

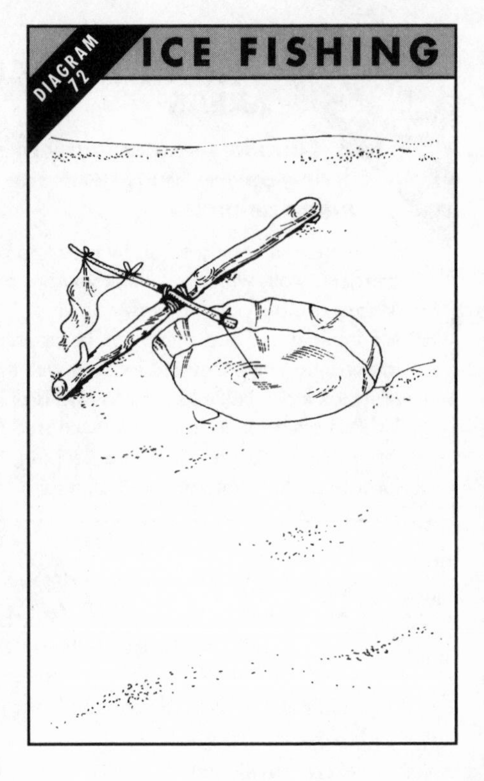

Ice fishing (Diagram 72) When lakes are frozen, fish usually congregate in the deepest water. Cut through the ice at this point and bait a hook. Make a pennant from cloth or paper and attach it to a light stick. Tie it firmly at right angles to another stick which is longer than the diameter of the hole in the ice. Fasten the fishing line to the other end of the flag pole and rest the pennant on the side of the hole. When a fish takes the bait the flag pole will be jerked upright. The following fish are available in snow and ice areas:

ARCTIC COD: can be caught in the ocean.
TOM COD: can be caught in the ocean.
SCULPIN: usually found in the ocean.
EELPOUT: can be caught in the ocean.
CRAWFISH: can be caught easily during the warm season.
SNAILS: like crawfish, can be caught during the warm season.
KING CRAB: this is one of the world's meatiest and tastiest crabs. In the spring it comes close to shore and can be caught on fish lines set in deep water, or by lowering baited lines through holes cut in the ice (see above).
SALMON: found in the freshwater rivers and lakes.
TROUT: also found in freshwater areas.
GRAYLING: a silver-grey freshwater fish, which resembles a salmon, that is present in large quantities in snow and ice areas.
Take precautions if you are using fish as a food source. Do not eat shellfish that are not covered at high tide. Never eat any shellfish that is dead when you find it, or any that do not close tightly when touched. The eggs of the salmon, herring and freshwater sturgeon are safe to eat, whereas those of the sculpin (which have large shiny heads) are not. In snow and ice areas, the black mussel

can be very poisonous. If mussels are the only food available, eat only those in deep inlets far out to sea. Remove the black meat and eat only the white meat.

Storing and preparing food In any survival situation you must take food when and where you find it. Do not wait. For example, you may stumble across a herd of caribou. In this situation you may think that you have time on your hands – you don't! Caribou are migratory animals and they are probably just passing through. When you wake up tomorrow they may all be gone and you will have lost a valuable food source. Therefore you must try to kill as many as possible (and this rule applies to any edible animals that you happen to come across). But now you have a problem: how will you stop the meat from spoiling?

As mentioned earlier, if it is winter then you can freeze the meat. In fact, your main problem will be getting the skin off as quickly as possible and cutting the meat up into sections. You should prepare and store food from animals in the following ways:

GAME: skin the animal (see Food Chapter). Cut the meat into small sections and save the liver, heart and tongue. *DO NOT* eat the livers of polar bears or seals: they contain lethal concentrations of vitamin A. Survivors have even got sick eating the livers of sled dogs that have been mainly eating seals.

After dividing the meat into sections you should build a cache to store it in. Build a platform off the ground and cover the meat placed on it with boughs to keep birds away. Do not build your cache in your camp. In the summer it may attract bears, and the last thing you want is a bear sniffing round your camp looking for food.

Unless you are going to eat the meat straight away, you will want to preserve it. Cooling preserves meat: try to keep your catch as cool as possible. Cut the meat into sections and hang them in a cool place out of the sun. The air will start to dry the outside of the meat. Start a number of smokey fires to smoke the meat. Do not let the fires heat the meat: just smoke it (do not use conifer wood – it ruins the flavour). Smoked meat, if kept cool and dry, can be stored in this condition for a long time.

Dry smoking preserves the meat even longer. Cut the meat up into long, thin strips and hang them on a drying rack. Dry the meat out in sunlight or with a fire. At the same time, have a smokey hardwood fire going When the meat seems dry move it nearer the fire for a few minutes to raise the temperature of the meat. When the strips are hot to the touch, move them back from the fire and continue to dry them until they become brittle. In this state they will last a long time. You can chew on them or cook them in water to rehydrate them.

Alternatively, you can make a brine solution and soak the meat strips in it before drying. To make a brine solution, fill a container with sea water and bring it to the boil. Add more sea water as the water in the container evaporates. Then fill the container to the top and let it cool. Do not use unboiled sea water for

brine: it does not contain enough salt to be sterile. The meat will obviously taste salty, but it will keep for long periods.
FISH: If you are near a water source, try to keep fish alive so that they are fresh when you want them. Pass a cord through the fish's lower jaw and tie it with a knot. Place the fish in the water and fasten the cord to a secure object on the shore. The fish will be able to breathe and stay alive for days. If you are not near water and you have to transport your fish or shellfish over long distances, keep them wet, cool and out of the sun. You can store dead fish on land for up to two days by building an evaporator: place the fish in a shady area and cover them with layers of wet grass and seaweed.

BUILDING SHELTERS

If you are in a polar region, it is very important that you get out of the wind and into some sort of shelter very quickly. If you make use of natural shelters, remember to avoid the lee side of cliffs where snow can drift and bury your shelter, and areas where avalanches or rock falls are likely.

Factors that determine what type of terrain you will build in are terrain, natural resources available, weather and snow conditions. *DO NOT* try to battle with nature: work in harmony with it. For example, if there are trees you can build an A-frame or lean-to shelter; if you are in the arctic above the tree line you will want to build a snow cave or snow trench.

If you are making shelters from snow blocks you will need a saw knife, snow knife, shovel or machete to cut snow blocks. The snow from which you cut the

SPETSNAZ TIPS

ARCTIC SHELTERS

Russia's deadly arctic warriors are experts at fighting and surviving in the world's coldest regions. These are their tips for when sheltering.

- Do not sleep on bare ground. Use insulating materials such as spruce or pine boughs, dry grass, dried moss or leaves.
- Do not cut wood that is over-sized for your shelter: it uses valuable energy and requires more cord for lashings.
- The superstructure poles must be the largest and strongest: everything else rests on them.
- Do not scatter your equipment on the ground: keep it in one place to avoid losing it.
- Have a fire going while you are building a shelter: it can be used as a heat source, a morale booster and can provide boiling water for you to drink later.
- Use clove hitches and finish with square knots for securing boughs together.

blocks should be firm enough to support your weight. Try to find a place where drifts are deep enough to allow you to cut blocks from a vertical face (it is less strenuous). It is well worth spending time finding proper snow of an even, firm structure, with no hollow or softer layers. The blocks should be around 45 x 50cm (18 x 20in) and 10-20cm (4-8 in) thick.

When building shelters in snow and ice areas, always the following points in mind:

- Never lay a tool down in the snow: you will lose it.
- Never hurry: if you do accidents and mistakes will happen.
- Work with nature, not against it.
- When building a shelter, drink as much water as possible. Dehydration is a killer.
- Expend as little energy as possible when building. Try to be as close as possible to the source of fuel for your fire.
- Take off clothes to compensate for the warming caused by physical activity when you are building. If you do not, your clothing will become soaked with sweat and you will risk hypothermia.
- Always take time to survey your surroundings when you are planning a shelter.
- For summer sheltering, remember that insects do not like wind, smoke and plants such as yew.
- Always protect yourself with spruce boughs or some other form of insulation from the cold and wet on the floor of your shelter.

Shelter living Regardless of the type of shelter you build in snow and ice areas, there are a number of principles you must follow in order to make your stay in them as comfortable as possible:

ENTRANCES: limit the number of entrances to your shelter to conserve heat. In addition, as fuel is generally scarce, keeping the entrances sealed will conserve your fuel supply.

ACTIVITY: if you have to go outside, remember to gather fuel, insulating material and snow or ice for melting. Do not waste your time.

WEAPONS: if you have a firearm, keeping it stored outside the shelter will prevent condensation gathering on it. Condensation could cause it to misfire.

LATRINES: try to relieve yourself indoors when possible. Try to dig connecting snow caves and use one as a toilet. Failing that, use tin cans for urinals and snow blocks for solid wastes.

INSULATION: always use thick insulation under yourself, even if you have a sleeping bag.

IMPROVISING: outer clothing makes good mattress material, the shirt and inner trousers can be rolled up and used as pillows.

SLEEPING BAGS: keep sleeping bags dry, clean and fluffed. To dry a sleeping bag, turn it inside out, beat out the frost and warm it in front of a fire. Be careful not to burn it.

SNOW: brush all snow off clothes before entering a shelter. Snow on clothing will melt inside a warm shelter and then will turn to ice when the clothing is taken outside. Remember: it is easier to keep clothing from getting wet than to dry it out later.

COLD: if you are cold during the night, exercise by fluttering your feet up and down or by beating the inside of your sleeping bag with your hands.

The shelters listed below will serve the survivor well in snow and ice regions. Remember to watch out for snow accumulating on the roof of your shelter – it may cause the roof to collapse.

Moulded dome shelter (Diagram 73) This is very quick to construct and requires minimum effort. However, you do need some sort of large cloth or poncho with which to build it. Pile up bark or boughs (not too large) and cover with material (A). Then cover the material with snow (remember to leave a gap for an entrance). When the snow has hardened, remove the cloth and brush (B). Make an entrance block from a number of small sticks wrapped inside a piece of cloth and then tied off (C), and remember to insulate the floor of the shelter with green boughs.

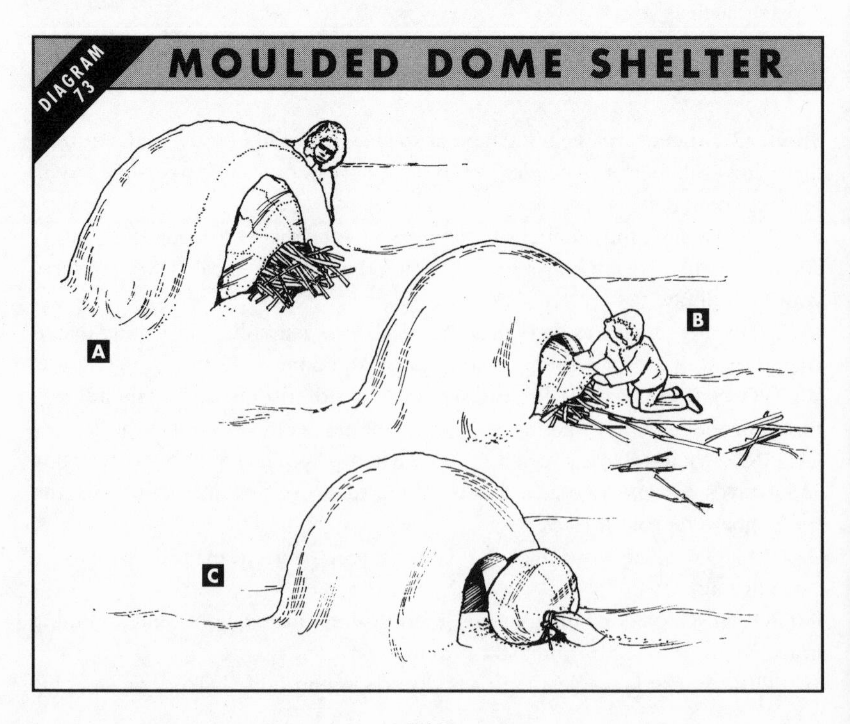

Snow cave (Diagram 74) A snow cave can be used in open areas where deep and compacted snow is available. Locate site on the lee side of a steep ridge or river bank where drifted snow gathers.

Choose the entrance carefully so that wind will not blow into the cave or drift snow block it. Burrow a small tunnel into the side of the drift for 1m (3ft) and then begin to excavate from this tunnel to the right and left, so that the length of the chamber is at right angles to the tunnel entrance.

The tunnel entrance must give access to the lowest level of the chamber (where cold air collects): where the cooking is done and equipment stored. You must ensure that the cave is high enough to sit up in. It should have an arched roof, partly for strength and partly to ensure that drops of water will run down the sides and not drip on the occupants.The sleeping area is on a higher level than the highest point of the tunnel entrance so that it will be in an area of the cave with warmer air (A).

The roof has to be at least 30mm (1.2in) thick and the entrance should be blocked with a rucksack, poncho or snow block to retain warmth. Remember to use ground insulation. The cave should have at least two ventilation holes on in the roof and one in the door. Be especially careful to keep the cave ventilated if you are cooking or heating inside it. Use boughs or some other form of insulation to protect occupants from the cold (B).

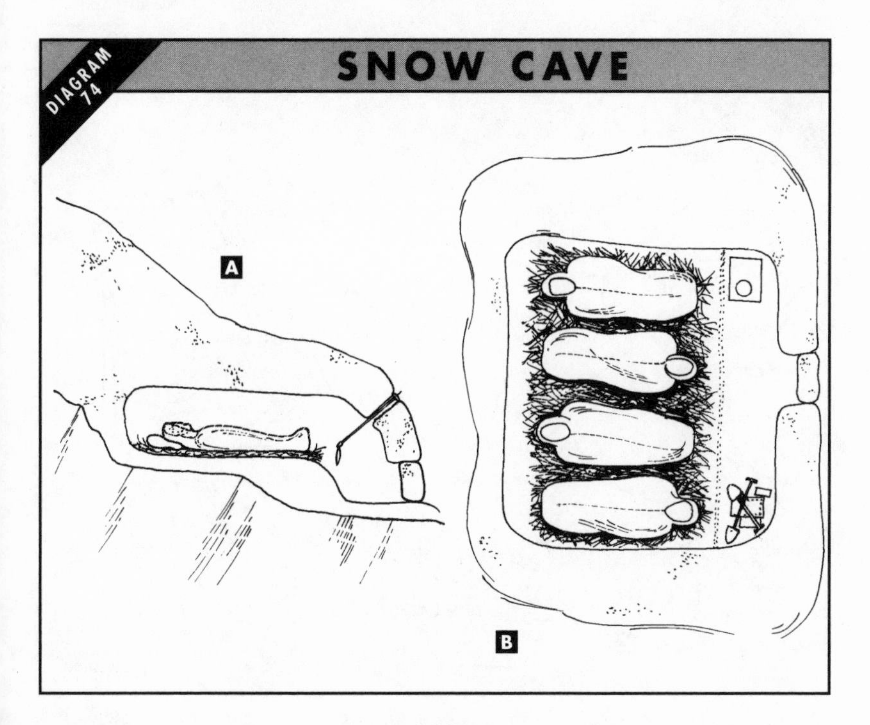

Trench shelter (Diagram 75) A temporary shelter that allows the survivor time to build a more permanent one. Find a large drift of snow at least 1m (3ft) deep and cut blocks to form a trench just wider than a sleeping bag and long enough to accommodate you (A). Build a wall of blocks around the trench (B) and roof over (C) with large slabs (hollow them slightly on the inside to form an arch). Don't forget a ventilation hole. The trench shelter should not be used long-term. If you are going to be in the area for a long time, build an igloo (see below).

Igloo (Diagram 76) Used by Eskimos for centuries, the igloo can be used as a long-term shelter in the polar regions. Draw a circle, 2.5-3m (7-10ft) on the snow to mark the inside diameter of the igloo. Cut snow blocks for the igloo from a trench nearby. When you have around 12 blocks begin to build the igloo. The first row slopes inwards with the end joints of each snow block having faces radial to the igloo centre.

Next, cut the spiral that will end at the key block. Cut right to left or left to right, whichever suits you (A). Begin next layer of blocks (B), and don't forget to bevel the tops of blocks to the igloo curves inwards (C). When fitting the key block (D), the hole should be longer than it is wide to permit passing the key block up through then juggling it into position. Then let it settle into position. Build on entrance tunnel to complete the igloo (E).

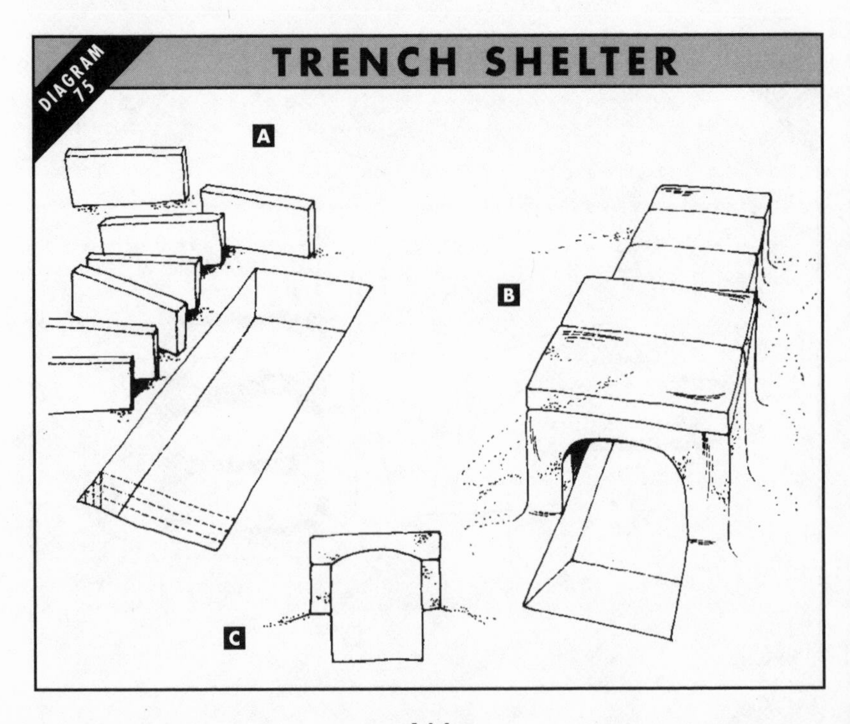

Inside the igloo you should have a tunnel entrance to trap cold air, a cooking level and a sleeping level (F). Put powdery snow on the dome and into open seams: it will harden and be an insulator. You may want to build a wind break around the igloo to prevent wind erosion. Don't forget to make ventilation holes and put insulating material on the sleeping level.

The cooking level must be reasonably close to the sleeping bench so anyone cooking does not have to rise from the bench. Be aware of high winds. If they are blowing they can cause drifting snow to erode the wall of the igloo. A snow wall should be built to act as a windbreak.

Inside the igloo, all sleeping bags should be placed side by side on the sleeping bench with their head end pointed towards the entrance. The igloo can be equipped with the following items:

STOVE: situate it near the entrance. Drive pegs into the wall above the stove and hang pots from them.

DRYING RACK: force sticks into the wall above the heat source. From these sticks you can dry clothing (always scrape snow off clothes before drying – never melt snow on garments) and thaw rations that do not need cooking.

DOOR: use a snow block for the door. Keep it open during the day and close it at night.

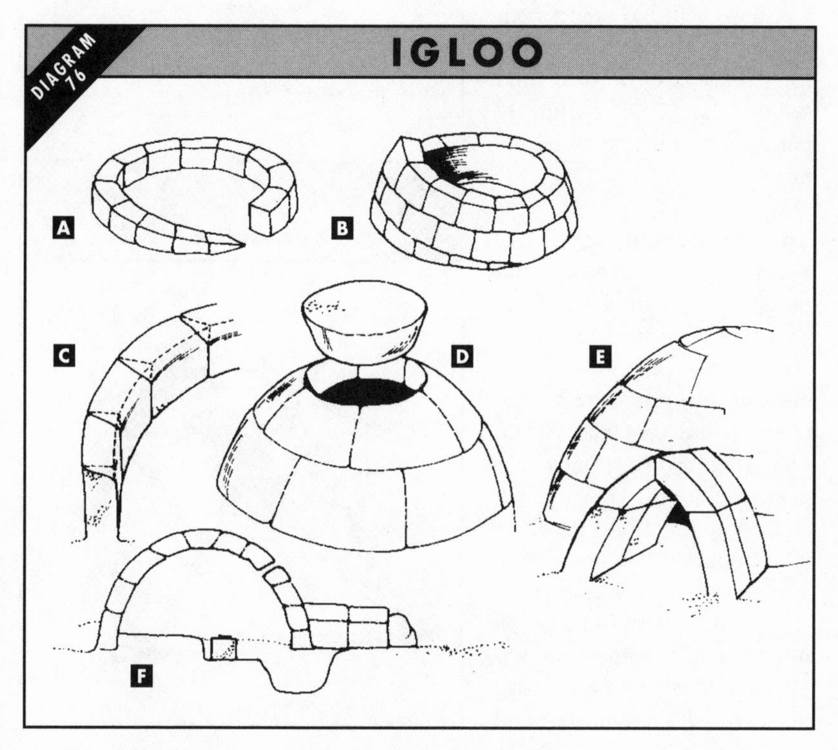

MINIATURE IGLOO: build a smaller igloo at the side of the main one if you find the igloo is becoming cluttered. Cut a doorway through the main igloo wall to form a cache.
Inside the igloo you must not prepare food by frying, baking or broiling. Heat canned goods in water.

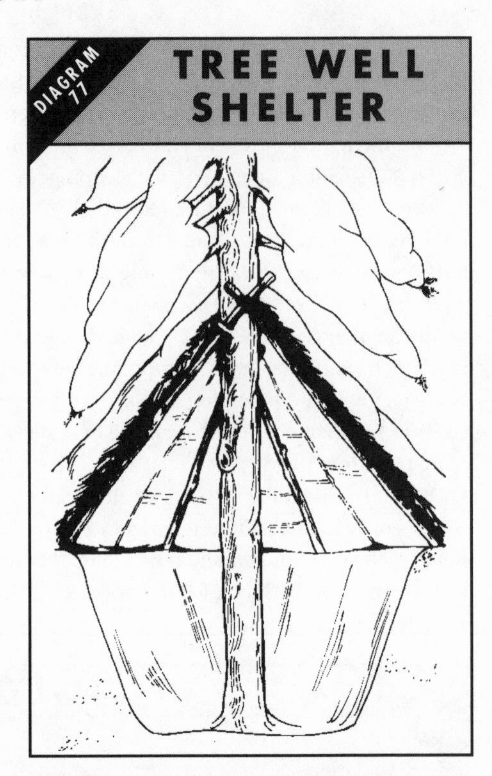

Tree well shelter (Diagram 77) Select a large tree which has thick lower branches and is surrounded by deep snow. Enlarge the natural pit around the tree formed by the snow. Make a roof with cut branches and boughs and line the walls and floor with the sam materials. This is a temporary shelter for use below the tree line. When making this shelter, try not to disturb the snow on the branches.

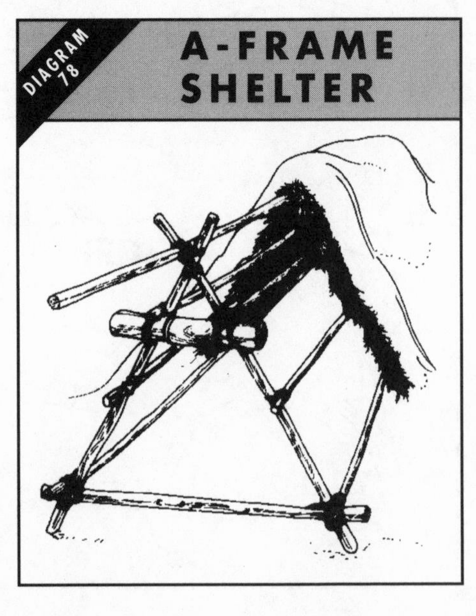

A-frame shelter (Diagram 78) A very simple shelter that can be constructed in a relatively short time. Ensure your superstructure poles are strong enough to support the weight of the shelter materials. When constructing the framework, position the poles horizontally and cover them and the floor with boughs. Then cover with snow and make a door plug.

To make this shelter you will need the following: one 3.5-5.5m (12-18ft) long sturdy ridge pole with all the branches and other projections cut off; two bipod poles approximately 2m (7ft) long; materials to go over the A-frame or cut branches to form a framework (as in the diagram); lashing material; and 14 stakes if you are going to use material over the A-frame.

Lash the two bipod poles together at eye-level height and place the ridge pole – with the large end on the ground – into the bipod formed by the poles and secure with a square lash (see Ropes and Knots Chapter). The bipod structure should be at a 90-degree angle to the ridge pole, with the bipod poles being spread out to an approximate angle of 60 degrees.

If you are using material to cover the A-frame, use clove hitches and half hitches to secure the fabric to the front of the ridge pole. Stake the fabric down with the stakes starting at the rear of the shelter and alternately staking from side to side to the front of the shelter. The stakes should be slanted or inclined away from the direction of pull. When you are tying off with a clove hitch, the line should pass in front of the stake first and then pass under itself.

When you are choosing poles for the framework, ensure that all the rough edges and stubs have been removed. This will ensure that you will not get injured when you are crawling in and out. If you are using natural materials for the covering, you should use the shingle method. Start at the bottom and work towards the top of the shelter, with the bottom of each piece overlapping the top of the preceding piece: this will allow water to drain off. Ensure that you use enough material to make a thick covering.

Lean-to shelter (Diagram 79)
This shelter is easy to make and can be both a summer or winter shelter. It will keep out insects, shield you from rain and snow, and keep you warm. A fire should be built directly in front of it with a fire reflector on the other side to reflect heat back into the lean-to. When you have built your framework cover it with boughs, starting from the bottom and working your way up with shingling, ie overlapping. Don't forget to insulate the floor (see above).

When you have constructed your shelter, add a door, build a fire reflector, a porch or a work

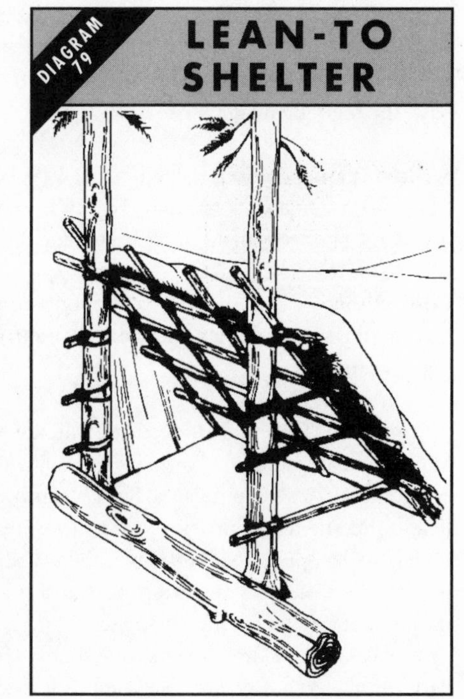

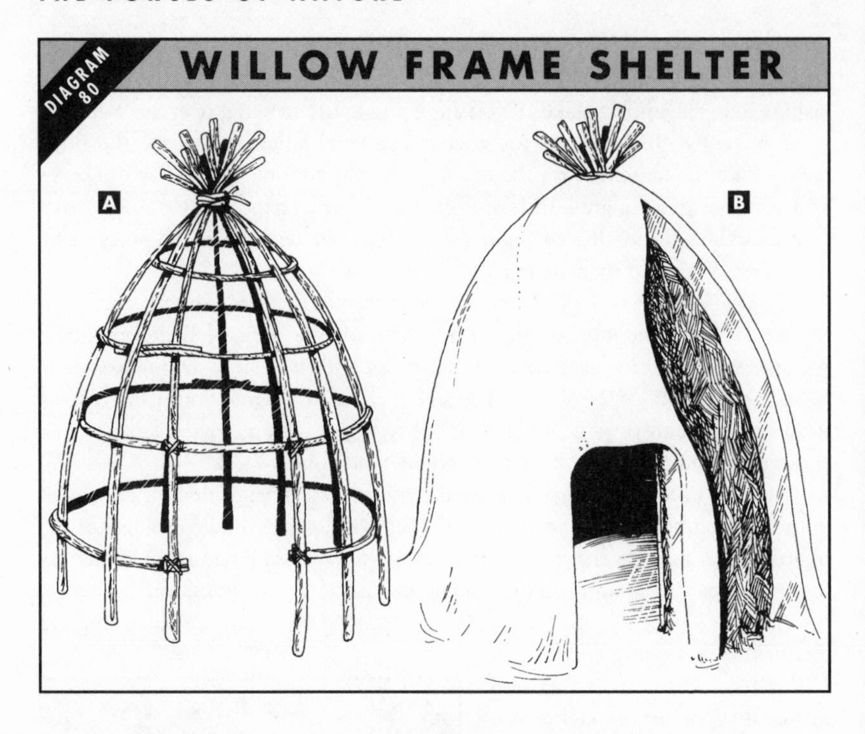

area. You can even build another lean-to to face the first one you built if you have the time. Be imaginative.

Willow frame shelter (Diagram 80) Very similar to the A-frame and lean-to. Construct a framework (A) and cover it from the bottom up with boughs. Cover the whole shelter with snow (B) in winter.

Cold climate camps Having a properly arranged camp can make your job of staying alive much easier and can even introduce a degree of comfort into the overall situation.

You should always build a fire reflector to direct heat from the fire back towards you. Line the inside of the shelter with foil blankets (if available) to improve heat retention. Cover your floor with spruce or pine boughs to give a comfortable, dry sleeping area. Stones can be heated by the fire and then placed inside the shelter to provide even more heat (put flat heated stones directly under the sleeping area and cover them with more boughs to stop you getting burnt). Try to keep your camp neat and tidy and compact: you don't want to have to venture far from your shelter.

Always ensure that you have sufficient fuel supplies for your fire. Decide early on what type of fire you want. For example, if you build a log cabin fire you will have lots of warmth and light, however it will burn quickly and therefore requires lots of fuel. In a snow area you could use up a lot of calories unnecessarily by replacing fuel supplies that should not have been used up in the first place. Choose a fire that burns for a long time with minimum fuel requirements (you can always add more fuel if you feel cold). See the Fire Chapter for the different types of fire you can build. Most importantly, do not let your fire go out: use an overnight fire.

Insects can be a problem when shelter-living in snow and ice areas, especially in summer. You may not even consider them when you are selecting a shelter site, but you should: they can make your life intolerable. Take preventative measures right from the start.

You may consider building a raised platform and coating each leg with oil or wiping them with yarrow (one of many plants that is a natural insect repellent). It has finely divided leaves and flat clusters of white flowers. In addition, you can rub yarrow on your skin to keep the midges off your skin.

As a general rule, stay away from deep woods and bodies of standing water. This is where insects thrive and are in their greatest numbers. Build your shelter where there is plenty of sunlight and breeze – there will not be as many insects. If you are really being pestered by them (and insects can make the survivor's life hell), make a fire or a number of fires and ensure there is always some smoke around you. You may not like it very much, but the insects will like it a lot less. Use small fires with green or rotten damp wood to guarantee plenty of insect-repelling smoke.

US ARMY TIPS

HEAT AND INSULATION FOR POLAR SHELTERS

US Army personnel are trained to fight in arctic regions, and part of their training covers the proper construction of shelters.

- Heat radiates from bare ground and from ice masses over water. Therefore, dig down to bare earth in shelter areas on land.
- If survivors cannot see their breath inside a shelter, it means the temperature inside the shelter is too warm. This will result in dripping and melting.
- To keep breath moisture from wetting sleeping bags, improvise a moisture cloth from a piece of clothing and wrap it around your head to trap your breath inside it.
- Once the inside of a shelter glazes over with ice, chip it off or build a new shelter – ice reduces the insulating qualities of a shelter.
- Check ventilation holes regularly.

MOUNTAINOUS TERRAIN

Mountains are hostile and dangerous. Freezing winds, driving snow, ice fields, mist, rain and sheer drops of hundreds of metres are all potential killers. The survivor must learn how to outflank them all to reach civilisation.

In normal circumstances, mountains and ice fields should only be climbed by experienced and properly equipped mountaineers. However, if you are stranded in mountainous terrain you must know how to get yourself out of danger and back to civilisation.

CHARACTERISTICS OF THE TERRAIN

On mountains you are likely to encounter snow, ice and winds, and may come into close close contact with avalanches of earth, rock and snow and crevasses. Weather conditions are erratic.

Avalanches Occur most commonly and frequently during winter, but also in warm temperatures and during the spring rains. They invariably have great force. The loose snow, or sluff, avalanche is one type that starts over a small area or in one specific spot, and grows as it gathers pace.

There are a number of factors that cause and affect avalanches and you should be aware of them. If you know where they are likely to strike, you can avoid being caught up in them.

STEEPNESS: avalanches occur most commonly on slopes ranging from 30-45 degrees. However, large avalanches can also occur on slopes with an angle of 25-60 degrees.

PROFILE: slab avalanches, which are the more dangerous kind, have a greater chance of occurring on convex slopes (Diagram 81) because of the angle and gravitational pull.

SLOPES: snow slides in mid-winter usually occur on north-facing slopes, which do not receive the sunlight required to stop the snow pack from getting very

US ARMY RANGERS TIPS

AVALANCHE WARNING SIGNS

As they operate in mountainous terrain, all US Rangers must have an in-depth knowledge of avalanches and where and when they occur.

- Avalanches usually occur in the same area. After a path has been smoothed it is easier for another slide to occur. Steep, open gullies, pushed over trees and tumbled rocks are signs of slide slopes.
- On leeward slopes snowballs tumbling downhill or sliding snow is an indication of an avalanche area.
- If snow sounds hollow there is danger of an avalanche.
- If snow cracks and cracks persist or run, a slab avalanche is imminent.

cold (snow stabilises better when the temperature is just above freezing). South-facing slope slides occur most frequently on sunny, spring days when sufficient warmth melts the snow crystals and changes them into wet, watery slides.

The leeward slopes of a mountain are hazardous because the wind blows snow into well-packed drifts just below the crest. If the surface snow is not attached to the snow underneath, a slab avalanche can occur.

Windward slopes generally have less snow and are more compact and are therefore usually strong enough to resist any movement. However, they too can be prone to avalanches when subjected to warm temperatures and moisture.

SURFACE FEATURES: avalanches are common on smooth, grassy slopes, ie those that offer little resistance. Trees and large rocks, on the other hand, can bind snow and prevent avalanches (Diagram 82).

OLD SNOW: old snow can cover up natural anchors, such as rocks, causing new snow to slide. However, an old snow surface that is rough and jagged will hold new snow much better than a smooth surface.

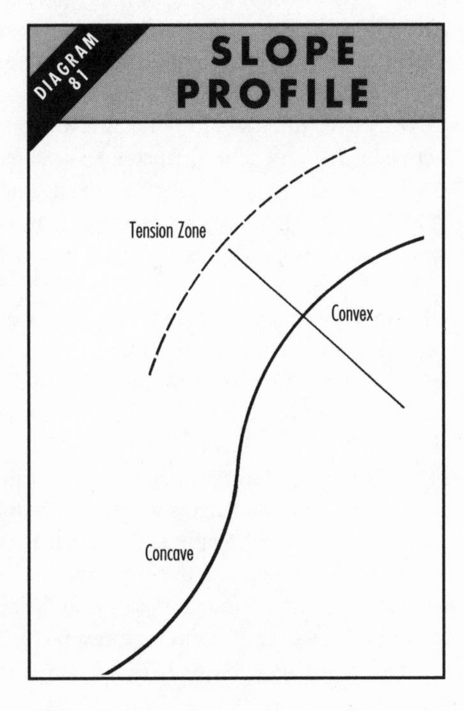

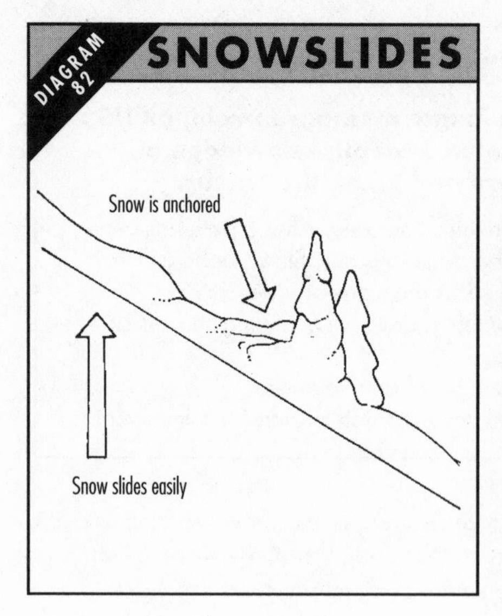

LOOSE SNOW: loose snow underneath compacted snow will make a snow slide more likely (there is no rough texture to restrain it). You should always check the underlying snow with a stick to see if it loose.
WINDS: a wind speed of over 25km/hr (15mph) increases the danger of an avalanche: leeward slopes will collect snow that has been blown from the windward sides.
STORMS: a high percentage of avalanches occur after storms.
SNOWFALL: a heavy snowfall over several days is not as dangerous as a heavy snowfall over a few hours. A slow accumulation allows the snow to settle and stabilise, whereas a heavy fall over a short time doesn't give the snow time to settle.
LIGHT SNOW: will not settle or bond and is therefore unstable.
DRY SNOW: has the same qualities as light snow.
COLD TEMPERATURES: under very cold temperatures snow is unstable. Around freezing point, or just above, snow tends to settle and stabilise quickly.
EXTREME TEMPERATURE DIFFERENCES: extremes of temperature, especially between day and night, cause adjustments and movement within the snow pack. You must keep alert for any quick temperature changes.
SPRING WEATHER: the sun, rain storms and warm temperatures associated with spring will cause avalanches, especially on south-facing slopes.

Glaciers These slow-moving masses of ice can pose a threat to the survivor. Glacial streams can run just under the surface of the snow or ice, creating weak spots, or they may run on the surface and cause slick ice. A glacier is essentially a river of ice that flows at a speed which depends largely on its mass and the slope of its bed. It has two parts: the lower glacier, which has an ice surface devoid of snow during the summer, and the upper glacier, which is covered with layers of accumulated snow that turns into glacier ice.

There are several features associated with glaciers, all of which you should know if you are forced to travel across them.
GLACIAL ICE: ice is smooth near the surface of a glacier, but not smooth enough to prevent cracking as it moves forward over irregularities in its bed. These fractures in the glacier surface are called crevasses (see below).

ROYAL MARINES TIPS

CAUGHT IN AN AVALANCHE

Britain's arctic warriors are experts in surviving in mountainous terrain. Follow their guidelines for what to do in an avalanche.

- Remove backpack and skis.
- Try to work towards the side of an avalanche.
- If swimming movements are possible, a double-action back stroke is the most effective, with your back to the force of the avalanche and the head up.
- Keep your mouth shut. In a powder snow avalanche, cover your mouth and nose with clothing to form an air space.
- Save your strength for when the avalanche loses momentum and settles.
- You must try to get to an air space near the surface, otherwise your chances of survival are minimal.
- Try to dig slowly to the surface.
- Do not panic.

ICE FALL: occurs where an abrupt steeping of slope happens in the course of a glacier. These ice falls consist of a mass of ice blocks and troughs, with many crevasses at irregular intervals.

LATERAL MORAINES: these occur along the receding margins of a glacier. When a glacier moves forward, debris(rocks and stones) from the valley on either side is dumped on its surface. When the glacier shrinks because of melting, this debris is deposited along its receding margins.

MEDIAL MORAINES: occur when two glaciers join and flow as a single river of ice. The debris also unites and flows with the major ice flow.

TERMINAL MORAINE: found where the frontage of the glacier has pushed forward as far as it can go, ie to the point at which the rate of melting at the front of the glacier equals the speed of advance of the ice mass.

GLACIAL RIVERS: vary in type. They can present problems if you have to cross them. Melting snow results in concentration of water pouring downwards in a series of falls and swift chutes. The sun's heat can release huge quantities of water from glaciers, resulting in a multitude of sub-glacial rivers and reservoirs forming under an ice field's surface. The level of glacial rivers will rise and fall: the flow usually eases at nightfall or dawn.

FLOODING GLACIERS: these are glaciers from which torrents of water flow. They are caused by the violent release of water that the glacier carried on its surface as lakes, or the violent release of large lakes that have been dammed up in tributary glaciers because of the blocking of the tributary valley by the main glacier. The water is released by a crevasse or a break in the moving glacial dam.

SWISS ALPINE TIPS

CREVASSE RESCUE

Switzerland's mountain troops are adept at crevasse rescue skills. Learn how to rescue climbers and survivors trapped in a crevasse.

- Pass a rope down with a loop in it. The suspended person can put a foot in it and therefore prevent their rope asphyxiating them.
- It takes three people to haul an unconscious person out of a crevasse – use manharness hitches.
- Temperatures in crevasses are very low, so speed is essential.

RUN-OFF CHANNELS: occur where melting takes place on the glacier. They cut deep channels in the ice and have smooth sides and undercut banks.Many of these streams end at the margins of the glacier, where in summer they contribute to the torrent that constantly flows between the ice and the lateral moraine.
GLACIAL MILLS: these are streams that disappear into the crevasses or round holes, where they then flow as sub-glacial streams. These glacial mills may, in some places, be as deep as the glacier itself.

Crevasses Found where a glacier starts at a valley wall, changes direction or spreads out in a winding valley. They vary in width from a few centimetres to hundreds of metres. They tend to be roughly parallel to each other in any given area, and they tend to develop across a slope. They can be covered with a thin layer of snow and are thus rendered invisible.

Snow-free slopes occur lower down the mountain, below the snow line. On these slopes you will encounter a variety of terrain, including scree, grassy slopes, forested areas and firm ground.

MOVEMENT

If there is no prospect of immediate rescue, you must get down into the valleys, towards civilisation and away from the cold and wet. Do not move in conditions of poor visibility or at night: you could injure yourself. Take time to survey the entire area around you. Look for a valley: it will probably be the beginning of a small stream or river. Select a safe route to get to it and find a way down. Do not go into avalanche chutes.

If you are on a high ridge stay away from overhangs: they may break off under your weight. Always try to walk down a spur ridge. The following points are very important for snow and ice mountains:

☐ Avalanches travel downhill: try to work out their most likely path.
☐ When you get to the main valley, stay away from possible avalanche run-outs.

- ☐ Always look above you for snow slides.
- ☐ You cannot outrun an avalanche.

Snow travel Try not to travel in thigh- or waist-deep snow: you will find it physically exhausting. South- and west-facing slopes offer hard surfaces late in the day after the surface has been exposed to the sun and then been refrozen. East- and north-facing slopes are generally soft and unstable. Slopes darkened by rocks or uprooted trees and vegetation provide more footing. Travel in the early morning after a cold night: snow conditions will be more stable then. Try to travel in shaded areas: the sun can make snow unstable.

If there are a number of you, travel in single file or in echelon formation. Try to walk around crevasses: it is generally much easier to do this than to try to force a crossing. A bride over a crevasse must be carefully examined. If snow obscures the bridge, the lead person must probe the immediate area closely. Be prepared for an arrest or sudden drop.

If the bridge is narrow or weak, a team can cross it by slithering on their stomachs, thereby lowering their centre of gravity and distributing the weight over a broader area. Where there is doubt about the bridge but it is the only route available, send across the lightest person first (ensure he or she is securely roped). Everyone should then follow, walking with light steps and taking care to step exactly in the same tracks.

Bridges vary in strength according to the temperature: in the cold of winter or early morning the thinnest and most fragile of bridges may have a very high structural strength. However, when the ice crystals melt in the warmer afternoon temperature even the most sturdy looking bridge may suddenly collapse.

BRITISH SAS TIPS

UPHILL AND DOWNHILL TRAVEL ACROSS SNOW

SAS soldiers are trained to fight in all types of terrain. Follow their rules for travelling up and down snow slopes on mountains.

- ■ Use zigzag routes to traverse steep slopes: it is less stressful than a straight uphill climb.
- ■ Always rope members of a team together for safety.
- ■ In a team, change the lead person frequently. Since he must choose the route of travel, he will get tired more quickly than the rest.
- ■ When traversing a snow plain, use the heels and not the toes to form a step.
- ■ When going downhill at speed, ensure all items of equipment, especially ice axes, are secured to your bergens.

If you decide to jump over a crevasse, you must adhere to these points:

- ☐ Decide whether you are going to make a standing jump or a running jump.
- ☐ Pack he snow down if you are planning a running jump.
- ☐ Locate the precise edge of the crevasse before jumping.
- ☐ Remove all bulky clothing and equipment before you jump.

When travelling up a snow slope, traversing (zigzagging) a slope is much easier than going straight up. In soft snow on steep slopes, pit steps must be stamped in for solid footing. On hard snow, however, where the surface is solid but slippery, level pit steps must be made. In both cases the steps are made by swinging the entire leg in towards the surface, not just by pushing your boot into the snow. In hard snow you may have to use crampons (spiked iron plates that clamp onto your boots) – which you should have if you are backpacker. Space steps evenly and close together to make travel easier and to retain your balance.

When descending a snow slope you can make use of the plunge step or step-by-step descending. The plunge step makes extensive use of the heels and can be used on scree (rock piles) as well as snow. The angle at which the heel should enter the surface varies with the surface hardness. On soft snow slopes, do not lean too far forward: you risk lodging your foot in a rut and suffering an injury. On hard snow, your heel will not penetrate the surface unless it has adequate force behind it. If you do not ensure your heel enters the snow, you may slip and begin to slide.

Step-by-step descending is used when the terrain is extremely steep, the snow very deep or you want to walk at a slower pace. You must face the slope and lower yourself step by step, thrusting the toe of each boot into the snow while maintaining an anchor with an ice axe.

Glissading If you are equipped with an ice axe, you can rapidly descend a slope by a method known as glissading. For the sitting glissade, you simply sit in the snow and slide down the slope using the ice axe as a brake. You can increase your speed by lying on your back to spread body weight and lifting your feet into the air. The standing glissade is similar to skiing: position yourself in a semi-crouch position with the knees bent as if sitting in a chair. The legs are spread outwards for stability, and one foot is advanced slightly to anticipate bumps and ruts. Speed can be increased by bringing the feet close together, reducing weight on the ice axe, and leaning forward until the boot soles are running flat along the surface like short skis.

When glissading, you must bear in mind the following points:

- ☐ Only make a glissade when there is a safe runout.
- ☐ Never attempt a glissade while wearing crampons: if they snag you could be thrown down the slope.
- ☐ You must wear mittens or gloves to protect your hands and keep control of the ice axe.

- ☐ Wear heavy waterproof trousers to protect your buttocks.
- ☐ Wear gaiters if you have them for a glissade.

Glacier travel If you are part of a group travelling across a glacier, remember to rope each other together. Travel in single file, stepping in the leader's footsteps. Exercise extreme caution when traversing a bridge over a crevasse. Use a stick or ice axe to probe the ice and snow.

Bear in mind the following points when you are travelling across glaciers:

- ☐ Lateral and medial moraines can offer excellent avenues of travel, especially if they are composed of large blocks.
- ☐ Moraine material that consists of small rocks, pebbles and earth will be loose and unstable.
- ☐ When a glacier is heavily crevassed, moraines may offer the only practical routes.
- ☐ When following a glacial river that is broken up into many shifting channels, select routes next to the bank. Do not run the risk of getting caught between two dangerous channels.
- ☐ Exercise extreme care when crossing a glacial surface stream: the bed and undercut banks are usually hard, smooth ice that offers no secure footing.

Snow and ice anchors If you are a backpacker, you should be equipped with the following items for travel over snow and ice mountains and glaciers. They will help you establish anchors or intermediate protection points during a climb or descent.

SNOW PICKETS: 0.9-1.2m (3-4ft) lengths of aluminium tubular sections that are used as long pitons and are suitable for belaying. Driven into the snow and then attached to the rope, they must always be used in pairs or greater numbers.

SNOW FLUKE: a large piece of metal that is literally driven into the snow surface at an angle. The softer the snow, the larger the plate should be. The fluke must be at a 45-degree angle to the surface of the slope for maximum effect. If placed securely, flukes will provide the same security as belay and rappel anchors (see below). Make sure you constantly check it for slippage.

TUBULAR ICE SCREWS: difficult to place in hard ice because they tend to clog and have a large cross section. Their main advantage is that they minimise 'spalling' (a crater-like splintering of the ice around the shaft of the screw) by allowing the displaced ice to work itself out through the core of the screw. This screw requires both hands for placement, though you can use an ice axe inserted in the eye for leverage.

COAT HANGER ICE SCREWS: they are thinner than ice screws and therefore easier to start in hard ice. Their holding power is less than tubular screws because they tend to fracture hard ice and, when under heavy strain, tend to shear because of their small cross section.

SOLID ICE SCREWS: developed in an attempt to have a screw that is easy to place and easy to remove. They offer good protection in soft ice but are less effective in other types of ice. Because of their limited thread displacement, melt-out is sometimes rapid. In addition, they tend to shear through ice when subjected to severe stress.

Rock and steep terrain You should have a rope for tackling rock faces, but if you don't then descend by facing the cliff on steep faces. On rock faces that are less steep, adopt a sideways position and use the inside hand for support.

When ascending move one foot and hand at a time. Make sure you have a good hold before continuing. Avoid becoming spread-eagled. Let your legs do the work. To climb up vertical fissures use the chimney technique: place your back against one surface and wedge your legs across the gap on the other. Move up slowly. Try to keep good balance when climbing: remember it is the feet, not the hands, that should carry the weight. Above all, avoid a spread-eagled position in which you stretch too far and then cannot let go.

The following holds will come in handy when you have to climb (always ensure you have a firm hold before you move):

PUSH HOLDS: (you literally push yourself away from a rock surface) help keep the climber low. Often used in combination with a pull hold.

PULL HOLDS: used to haul yourself up. They are the easiest holds to use.

ROYAL MARINES TIPS

PRINCIPLES OF MOVEMENT ON ROCK OR STEEP TERRAIN

In mountainous terrain you must know how to walk correctly. Remember you are in a hostile environment – follow Royal Marines guidelines.

- Conserve energy: always keep the centre of gravity over your feet to make the legs do most of the work, not the arms and upper body.
- Always test holds by tapping the rock and listening for a hollow sound, which indicates instability.
- Keep hands at shoulder level to ensure blood supply to arms and hands is not reduced.
- Watch where you put your feet.
- Keep three points of contact with the rock at all times.
- Carry out slow rhythmic movements.
- Think ahead: plan moves and anticipate any difficulties.
- Keep your heels down.
- Remove rings before you start climbing: fingers have been lost becuase rings have jammed in cracks.

US ARMY RANGERS TIPS

GENERAL PRECAUTIONS FOR MOUNTAIN WALKING

In mountainous terrain, US Rangers follow some very simple rules. Take note of them, and follow the example of America's crack troops.

- Do not kick rocks loose lest they roll downhill – they can be extremely dangerous to anyone below.
- Step over obstacles like rocks and fallen logs to avoid fatigue.
- Do not jump in mountains. Landing areas are invariably small, uneven and have loose rocks or dirt: you may slip and fall farther than intended.

JAM HOLDS: involve jamming any part of the body or extremity into a crack. Place your hand into a crack and clench it into a fist, or thrust your arm into a crack and twist your elbow against one side and your hand against the other. If you are using your foot to make a jam hold, remember to ensure that you can remove it easily when you want to climb on.

COUNTERFORCE HOLD: achieved by pinching a protruding rock between the thumb and fingers and pulling outwards, or by pressing inwards with the arms.

LAY-BACK HOLD: achieved by leaning to one side while your hands are in a crack in the rock, then pulling up with them. Your feet push against the rock, ie the hands and feet pull and push in opposite directions.

MANTLING: takes advantage of down pressure exerted by one or both hands on a slab or shelf. By placing your hands on the shelf, hauling yourself up and then straightening or locking your arms, your body is raised up and you can place a leg on a higher hold. This hold is also called a mantle shelf.

DIAGRAM 83

LOCKING KNEES

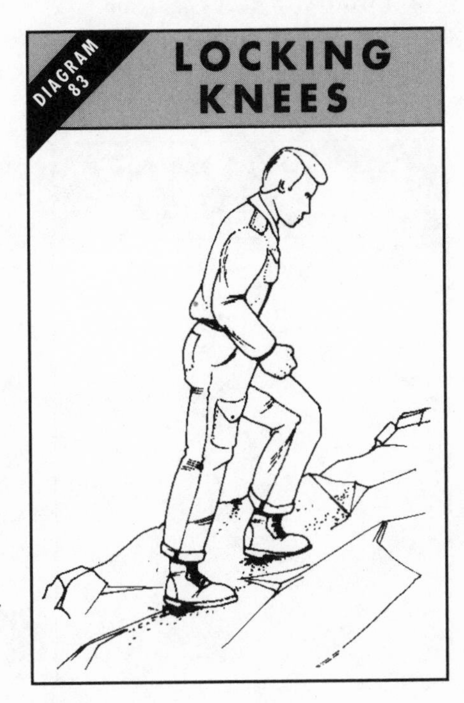

Mountain walking techniques

Below the snow line, regardless of the type of slope you are on, remember two points:

☐ Keep the weight of your body over the feet.
☐ The sole of your boot must be placed flat on the ground.
Take small steps at a steady pace. When ascending on hard ground, lock your knees with every step to rest the leg muscles (Diagram 83). If you encounter steep slopes, remember that traversing is easier than going straight up (Diagram 84). Turning at the end of each traverse is done by stepping off in the new direction with the uphill foot. This stops having to cross the feet and risking loss of balance. Take frequent rest stops: you make mistakes when you get tired, which can result in twisted ankles and broken legs.

For narrow stretches of uphill travel use the herringbone step (Diagram 85), ie ascending with the toes pointed out. When descending, keep your back straight and knees bent, with the weight kept directly over the feet.

On grassy slopes (Diagram 86A) step on the upper side of each tussock, where the ground is more level than on the lower side. When descending, it is best to traverse. Scree slopes are are made up of small rocks and gravel which have collected below rock ridges and cliffs. Ascending such slopes is difficult and potentially dangerous: kick in with the toe of your upper foot to form a step in the scree (Diagram 86B). When descending walk down the slope with your feet in a slightly pigeon-toed position using a short step. Go at a slow pace. On rocky slopes (Diagram 86C) step on top and on the uphill side of the rocks.

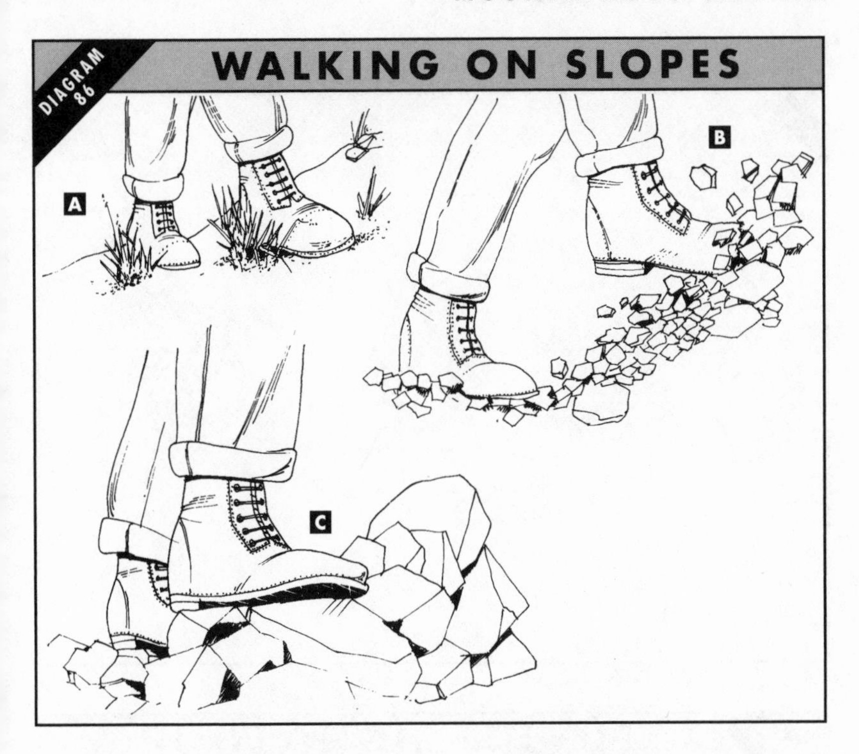

KNOTS

There are a number of specialised knots for climbing, and the survivor should be aware of them. They are designed to have the least effect on the fibre of a rope lock without slipping, and to be easy to untie when conditions are wet and icy. Though all knots reduce the strength of ropes, these knots are designed to reduce their strength as little as possible. Most knots should be made safe with an overhand knot or two half hitches (see Ropes and Knots Chapter). However, a knot does not have to be made safe if it is in the middle of a line.

Water knot (Diagram 87) Used for joining nylon webbing, not rope (also called the right bend).

Figure-eight loop (Diagram 88) The figure-eight loop can be tied at the end or in the middle of a line establishing a fixed loop. If the loop is tied at the end of the line, then an overhand or single fisherman's safety knot must be used.

Manharness hitch (Diagram 89) Also called the Butterfly knot, it is used to make a fixed loop in the middle of a line. Make a loop in the rope (A) – allow the left side of the loop to cross over the loop (B) – twist the loop (C) and pass it

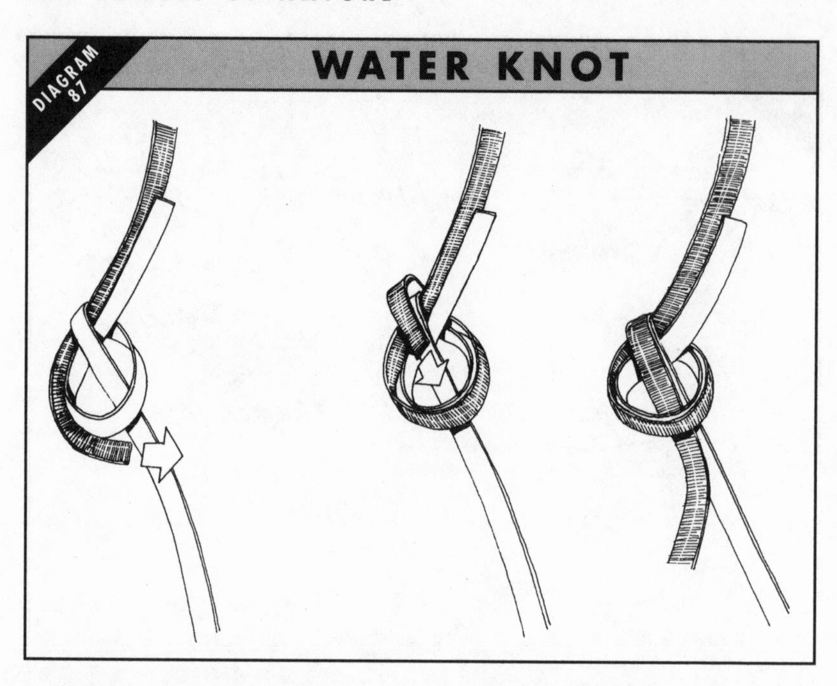
DIAGRAM 87
WATER KNOT

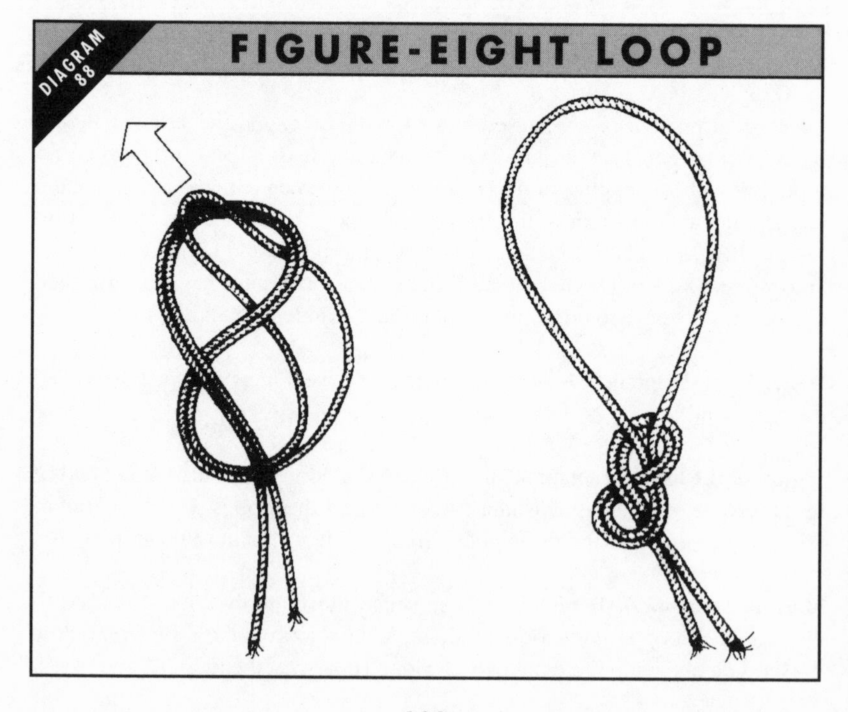
DIAGRAM 88
FIGURE-EIGHT LOOP

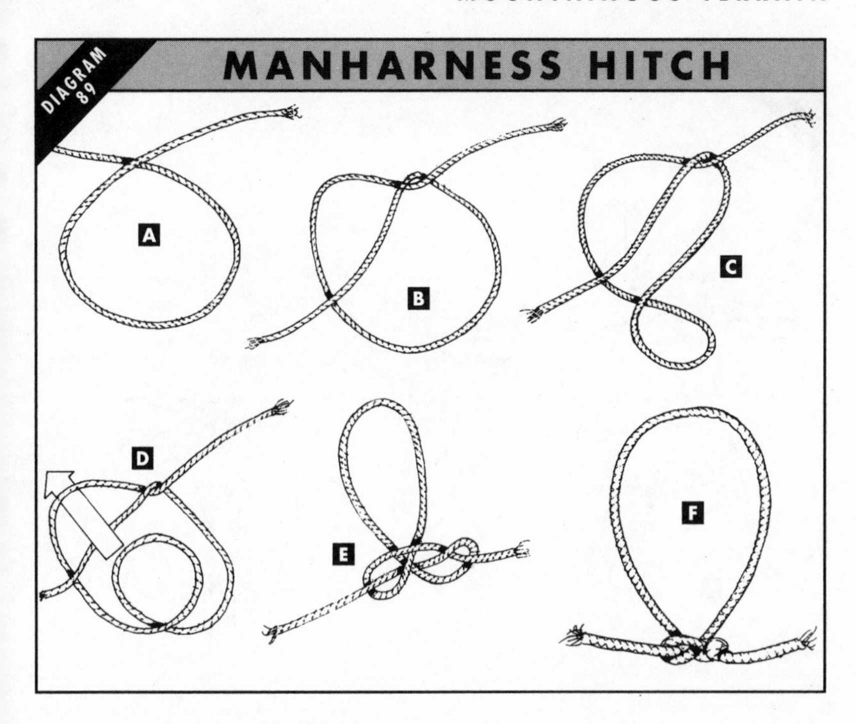

over the left part of the rope and through the upper part of the original loop (D). Pull knot gently into shape and tighten it (E). If you do not twist the loop, the final strength of the loop does not appear to be reduced. The end result of not twisting the loop is shown in (F).

Improvised seat harness (Diagram 90) Can be made from tubular nylon tape. Place tape across the back so that the hindpoint is on the hip opposite the hand that will be used for braking during belaying or rappelling. Keep the midpoint on the appropriate hip, cross the ends of the tape in front of the body and tie three or four overhand wraps where the tapes cross (A). From front to rear, bring the ends of the tape between the legs, around the legs and then secure with a hitch on both sides of the waist. Tighten the tapes by pulling down on the running ends to prevent them from crossing between the legs (B).

Bring both ends around to the front and across the tape again. Then bring the tape to the opposite side of the intended brake hand and tie a square knot with an overhaul knot or two half-hitch safety knots on either side of the square knot (see Ropes and Knots Chapter). The safety knots should be passed around as much of the tape as possible (C). Clip a carabinier (coupling link with safety closure) to the harness by clipping all the web around the waist and the web of the overhand wraps together.

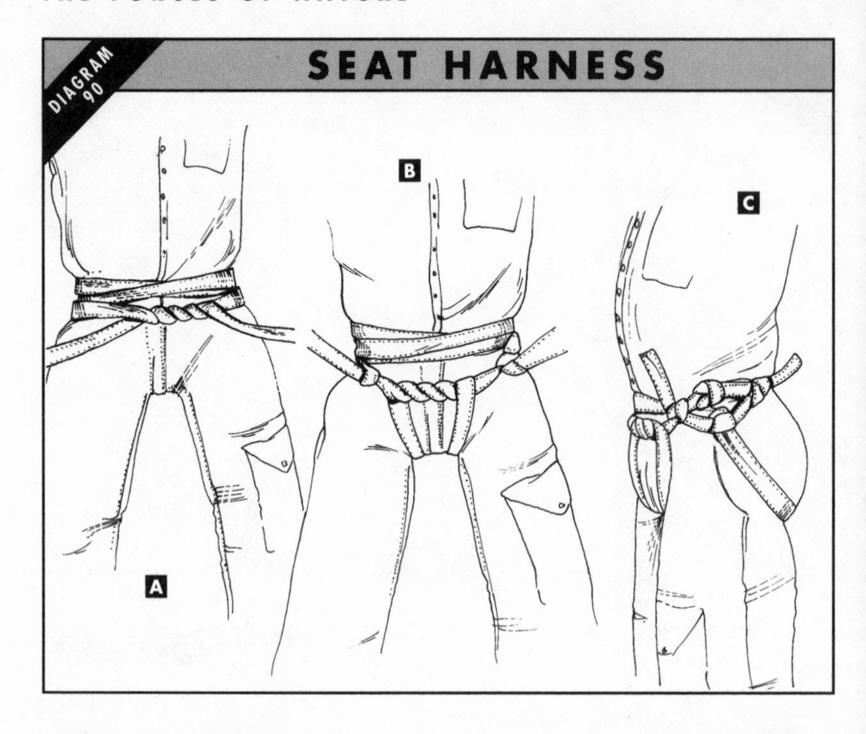

Belaying (Diagram 91) is a way of ascending for two or more people with ropes. One person (the climber) ascends with a rope attached around the waist with a bowline, while the belayer secures his ascent. The belayer anchors the rope with a loop tied in a figure-of-eight and ties on with a bight or two bights to steady himself. He passes the climbing rope over the head and down to the

US ARMY RANGERS TIPS

PROCEDURE FOR BELAYING

Rangers are highly trained in belaying and rope work – they need to be able to infiltrate enemy mountain positions quickly and speedily.

- Run rope through guiding knot and around the body.
- Anchor yourself to the rock with a portion of the climbing rope or a sling rope if your position is unsteady.
- Ensure remainder of the rope is laid out so as to run freely through the braking knot.
- Do not let too much slack develop and do not take up slack too suddenly – you could throw a climber off balance.
- In the event of a fall, relax the guiding hand, let the rope slide enough so that the braking action is applied gradually. Then hold belay position.

hips, making a twist around the arm closest to the anchor and takes up the slack. The climber ties on with a bowline around the waist and starts to ascend (A); the belayer takes in rope to keep it taught. It is important for the anchor belayer and climber to be in a straight line (B).

The sitting belay is the best position. The belayer sits and tries to get good bracing between his legs and buttocks. The legs should be straight, knees locked with the rope running around the hips.

Note that walking rope will only take a loading (breaking strain) of 907kg (2000lb), which is only just enough to save a man on a modest fall. A climber weighing 82kg (180lb) falling 30m (82ft) will, when he is brought up, exert an equivalent force on the rope of 1038kg (2288lb). A rope intended to protect climbers against vertical falls should have a breaking strain of 1900kg (4200lb). Failing this, use the double rope technique: two ropes together.

All climbers should be aware of the following dangers:

WET OR ICY ROCK: can make an easy route impassable.

SNOW: may cover holds.

SMOOTH ROCK SLABS: can be dangerous, especially if wet or icy.

ROCKS OVERGROWN WITH MOSS OR GRASS: treacherous when wet.

TUFTS OF GRASS OR SMALL BUSHES: may be growing from loosely packed and unanchored soil.

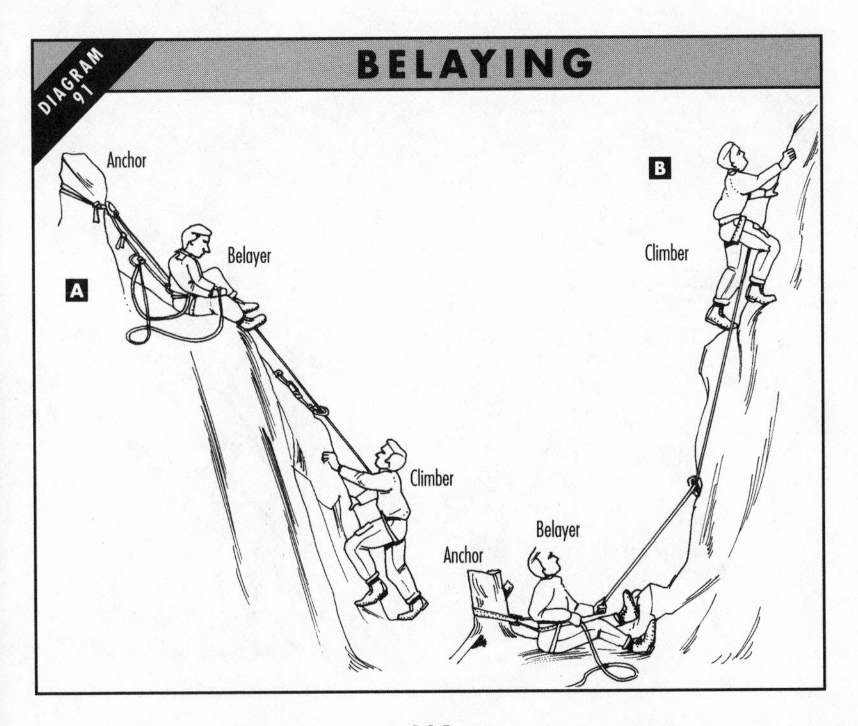

TALUS SLOPES: can be dangerous because of the threat of falling rocks.
ROCK FALLS: frequently caused by other climbers, heavy rain and extreme temperature changes in high mountains. In the event of a rock fall seek shelter, or, if this is not possible, lean into the slope to minimise exposure.
RIDGES: can be topped with unstable blocks.

Rappelling (Diagram 92) A survivor with a rope can descend quickly by sliding down a rope that has been doubled around an anchor point. When rappelling, ensure that the rope reaches the bottom or a place from where further rappels can be made.

The rappel point should be carefully tested and inspected to ensure the rope will run around it when one end is pulled from below. Make sure too that the area is clear of loose rocks, otherwise the rope may dislodge them during the rappel. If this happens, they may fall on persons below and inflict injuries.

The climber faces the anchor point and straddles the rope, then pulls it from behind, runs it around either hip, diagonally across the chest, and back over the opposite shoulder (A). From there the rope runs to the braking hand, which is on the same side of the hip that the rope crosses. You must lean with the braking hand down and be facing slightly sideways (B). The foot on the same side as the braking hand should precede the other at all times.

ITALIAN ALPINE TROOPS TIPS

RAPPELLING TECHNIQUES

Italy's mountain troops are among the most skilled alpine soldiers in the world. Use their techniques for rappelling.

- Lean out at a 45-degree angle to the rock.
- Keep your legs well spread and straight for stability.
- Turn up your collar to prevent rope burn on the back of the neck.
- Wear gloves and other articles of clothing to pad the hands, shoulders and buttocks.
- To brake lean back and face directly into the rock so that the feet are flat on the rock.
- Keep feet shoulder width apart.

Anchors (Diagram 93) Some anchor systems are simple and consist of a single anchor point. Alternatively, they may be complex and made up of multiple anchor points. They provide protection for both the belayer and climber.

The basis for any type of anchor is strong, secure points for attachment. Natural anchor points include the following:

CHOCK STONE (A): a natural chock stone is a securely wedged stone that provides an anchor point for a sling. In most cases the rock is wedged within a crack.

BOLLARD (B): a rock bollard is a large rock or portion of rock that has an angular surface, over which a sling or rope can be placed in such a way that it will not slip off. You must take care to ensure that the bollard will not be pulled loose if it is subjected to a sudden load.

TREE ANCHOR (C): trees can make very secure anchor points, though in rocky or loose soil they should be avoided if other anchor points are available. If not, then you must watch the tree very carefully for slippage.

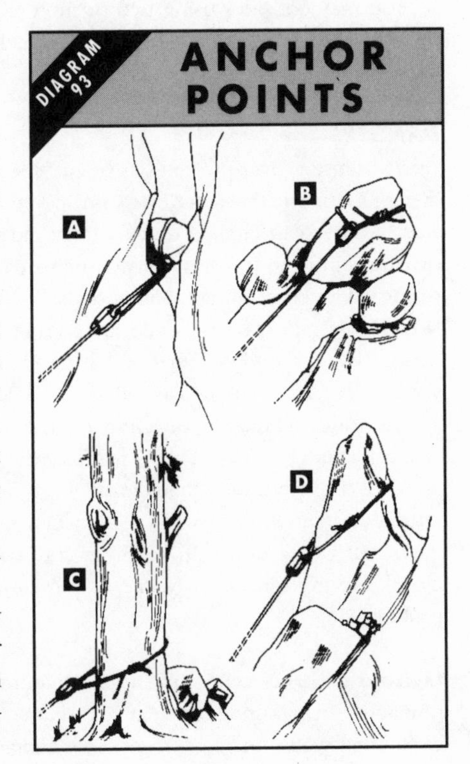

BRITISH SAS TIPS

ICE AND SNOW BOLLARDS

If there are no rocks or trees to use as bollards for rope work, do what the SAS does and fashion them out of ice or snow.

- Cut an anchor from the ice in the shape of a mushroom.
- Ensure diameter is a least 40cm (18in) and depth at least 15cm (6in).
- If ice cracks stop and make another bollard.
- Snow bollards must be at least 30cm (1ft) deep and from 1m (40in) wide in hard snow to 3m (10ft) in soft.
- Pack bergens around the bollard to prevent the rope cutting through it.

SPIKE (D): a spike is a vertical projection of rock. To use as an anchor point, place a sling around the spike.

As well as natural anchors, you can use artificial anchors: chocks (metal wedges that fit into cracks) and pitons (metal spikes that can be driven into a crevice to secure a rope).

You may consider using two or more anchor points, which will strengthen the whole anchor system. The one drawback with this is that if one anchor fails, the remaining points will be shock loaded.

DANGERS

Apart from the ever-present dangers of the cold and the wet, the sheer unpredictability of weather conditions on mountains can pose a number of hazards. The weather can change quickly from pleasant sunshine to gloomy skies and driving rain or snow storms. Mountains cause air currents to be uplifted and disturbed and they attract long periods of severe weather. The wind blows most strongly on mountain tops and across ridges because its speed increases with height. Do not underestimate a strong wind or its chilling effect: it will drain your energy as you try to stay balanced while being buffeted.

The windchill factor (see Surviving in Polar Regions) is nearly always present in mountainous terrain. Even on sunny days do not underestimate the chilling effect of the wind. Rain tends to be more frequent and heavier in mountainous terrain and can soak you to the skin in a short space of time: be aware of the danger of hypothermia. In addition, you will face reduced visibility due to low cloud, driving rain, mist, whiteouts or storms, all of which cause navigation problems.

Lightning This is very dangerous in mountains. It is attracted to summits and pinnacles. In a thunderstorm avoid summits, exposed ridges, pinnacles, gullies containing water and lone trees. Overhangs and recesses in cliffs do not offer protection against a discharging current. Leave wet ropes and metal equipment at least 15m (50ft) from your shelter if possible. Avoid vertical cliffs: they are excellent conductors. Adopt a sitting position with the knees drawn up against the chest. This is the best protection against earth currents.

Weather forecasts If you have access to a radio, listen to the general forecasts, or, better still, ring the local meteorological office for a personal weather service. It is very important to give the exact area and time required for the forecast. You should ask for information about the valley and mountain top weather, temperatures, winds, type of precipitation, visibility, likelihood of a freeze and any rapid changes approaching. Obviously if you are survivor you won't be able to do this, but there is really no excuse for the backpacker to be caught unawares.

Another thing to remember about the weather is that there will be a great deal of local knowledge about it. Shepherds, farmers, foresters and mountain rescue teams will all know their local area well, perhaps even better than a meteorological office. Always seek their advice before going onto high ground.

ROYAL MARINES TIPS

USING AN ALTIMETER AS A BAROMETER

If you have an altimeter you can use it to predict changes in the weather. In mountainous terrain such information can save your life. Note these Royal Marines guidelines:

- An appreciable rise in pressure over a few hours means the good weather will be short-lived.
- A rapid rise in pressure during a 24-hour period indicates good weather. However, if the rise lasts only a day, then the good weather will not last much longer.
- A slow and uniform pressure rise over 2-3 days indicates a long spell of dry weather
- A rise in pressure accompanied by wind veering from south through west to north means the weather will improve.
- Fog or mist is expected when the pressure goes up to an unusually high level with a moist atmosphere during a period of calm conditions.
- Unsettled weather is indicated by a rapid, erratic rise in pressure interspersed with small falls.
- Rain is indicated by the wind veering from north or east to south or southwest with accompanying falling pressure.
- A long continuous drop in pressure indicates a long period of rain.
- In calm conditions a rapid but brief fall in pressure, combined with increasing humidity, indicates a thunderstorm is coming.

INTERNATIONAL DISTRESS SIGNAL

On mountains, the recognised distress signal is six blasts with a whistle (or similar sound) repeated at one minute intervals. At night use six torchlight flashes. The reply is three blasts or flashes at one minute intervals.

FINDING WATER AND FOOD

There is little food on mountains, and none at all on high cliffs. You may encounter mountain goats and sheep on the lower slopes, but they are wary and difficult to approach. However, they can be surprised by moving quietly downwind when they are feeding with their heads lowered (remember that they are sure of their footing and you may not be – do not get injured chasing after a mountain goat). There may also be edible plants on the lower slopes (see Food Chapter). However, your first priority will be to get into the valleys, where there will be ample food.

Water is less of a problem on high ground: melted snow, ice and rain water collected directly can be drunk without purification.

BUILDING SHELTERS

Because mountain areas are predominantly rock, snow and ice (on the higher slopes at least) there will be few materials available to you for building shelters. Your best bet is to dig into the snow or ice if you have some sort of tool. Build a snow cave if you have some kind of cutting implement (see Polar Regions Chapter). Do not spend a lot of time trying to make shelters in mountainous terrain. Above all, remember that your number one priority is to get out of the wind. After that, you should get yourself off the mountain and down into the valleys as quickly as possible.

If you do have to make a shelter on a mountain, ensure you also build a fire to keep you warm. Use any material that is available: sticks, grasses, even dried goat dung if you can find it. Basically anything that will burn. Use fire reflectors if possible (see Fire Chapter).

BRITISH SAS TIPS

SHELTERS AND SLEEPING IN MOUNTAINS

Trapped on a mountain? Use the hard-earned experience of the SAS to help you with shelters and sleeping on rocky ground.

- Dig into the snow if there is no shelter among the rocks.
- A plastic bag can make an improvised sleeping bag.
- On a slope, sleep with your head uphill.
- On rough and stony ground, sleep on your stomach for greater comfort.

SURVIVING IN THE DESERT

Intense heat, lack of water and an apparent absence of food are only some of the problems you will face in the desert – but don't despair: with proper guidance you will be able to survive easily and get back to civilisation.

Deserts occupy around 20 per cent of the earth's land surface. However, the idea that they are all composed of sand is a misconception. There are in fact six types of desert: alkali, sand, rock, rocky plateau and mountain.

THE WORLD'S DESERTS

There are eight main desert areas in the world, each having its own set of distinct characteristics.

Sahara Desert Situated in North Africa, it has little vegetation and is characterised by loose, shifting sand. It has areas of sandstone, limestone and volcanic rock, salt marshes, canyons and marshes. It is swept by hot, dry winds that cause major sandstorms. The nights are bitterly cold during the winter months, often requiring the wearing of overcoats and blankets.

Arabian Desert Extends over 1,600,000km (1,000,000 miles) in the Middle East. It is the desert that can come close to being a complete wasteland: continuous, drifting dunes and almost no vegetation.

Persian Desert Extends from the Persian Gulf to the Caspian Sea. Its climate very is severe: the wind blows constantly from the north during the 'wind of 100 days' during the summer, with speeds of up to 120km/hr (75mph).

Gobi Desert Situated in China, it is a huge, waterless area of some 960,000 square kilometres (600,000 square miles) surrounded by high mountain barriers that prevent rainfall. Almost treeless, the Gobi is covered with wiry tufted grass.

Atacama Desert Situated in South America and covering areas of Chile, Argentina and Bolivia, this desert is characterised by an almost total absence of rainfall. It is completely barren and desolate, but is inhabited because it contains valuable mining deposits.

The Great Basin Situated in the United States, in the states of Arizona, New Mexico, Nevada, Utah, Texas, Colorado, California and northern Mexico, its terrain is rocky and contains ravines, canyons and escarpments. The vegetation consists of cactus and sage brush.

Kalahari Desert Located in the South African highlands, it consists of extensive areas of red sand and flats. A lot of the Kalahari is covered by a heavy growth of scrub trees.

Australian deserts The deserts of the Australian 'outback' are more or less uninhabited (apart from the Aborigines). They are characterised by erratic rainfall, cyclones and storms. Mostly flat, these areas contain only scrub growth.

CHARACTERISTICS OF THE TERRAIN

Though there are different types of desert areas, they all share common physical characteristics. It is important that you know what they are so you can prepare your survival plan.

Lack of water Deserts are characterised by an absence of water, which is why their human populations are low. Annual rainfall can be zero to 25cm (10in), but whatever rainfall there is is completely unpredictable. As a result, flash flooding (where normally dry stream beds are filled with quick-moving flows of water) is very common.

CANADIAN AIR FORCE TIPS

IMMEDIATE ACTIONS FOR THE SURVIVOR IN THE DESERT

Deserts are harsh environments – your immediate actions in the aftermath of a crash or being stranded are crucial. Follow Canadian Air Force guidelines:

- Do not walk blindly into the desert.
- Get into shade as quickly as possible.
- Keep your head and the back of your neck covered.
- Evaluate the situation calmly and then decide on a course of action.
- Your immediate priorities are administering first aid if you are injured, finding shelter and water.

Lack of vegetation Vegetation is generally scarce, and what plant life there is will be specially adapted to withstand the severity of desert conditions. The types present are an indication of the depth of the water table. Thus palm trees indicate water within 0.6-0.9m (2-3ft) of the surface; cottonwood and willow trees suggest it may be found 3-3.6m (10-12ft) from the surface. Note: the common sage, greasewood and cactus have no bearing on the water level and are thus useless as indicators of the level of the water table.

Temperature extremes Desert temperatures vary according to latitude. For example, the Gobi Desert experiences temperatures of -10 degrees C (-50 degrees F) in the winter. On the other hand, the Sahara Desert has recorded temperatures of up to 58 degrees C (136 degrees F). Because of the unobstructed, direct effect of the sun's rays during the day, temperatures are high, but at night they fall rapidly, especially on elevated plateaus, as the surface cools quickly under the clear night skies.

Bright sunshine and moonlight Because of the low cloud density, the days are abnormally bright and the nights crystal clear.

Dust storms Winds in the desert can reach hurricane force, throwing up dense clouds of dust and sand. As well as being extremely uncomfortable physically, visibility is reduced to almost zero.

Mirages The result of light refraction through heated air rising from very hot sandy or stone surfaces. They usually occur when you are looking towards the sun, and tend to distort the shape of objects, especially vertically. You may 'see' hills, mountains and lakes during your journey that are actually mirages.

Other desert terrain characteristics include hillocks, wadis (dry river beds and valleys) and oases.

Man-made features All deserts contain at least some man-made features. As a survivor, you should look out for them: they may lead to civilisation (though the distances may be great). In particular, watch out for the following:

ROADS AND TRAILS: most road systems have existed for centuries to connect centres of commerce or important religious shrines. In addition, there are often rudimentary trails for caravans and nomadic tribesmen, and these often have wells or oases every 32km (20 miles) or 64km (40 miles), though in some areas there may be over 160km (100 miles) between watering places.

BUILDINGS: in the desert, most structures for human habitation are thick-walled and have small windows. The ruins of earlier civilisations litter deserts – they can be used as temporary shelters on your journey.

PIPELINES: they can lead you to rescue, and as they are often elevated above the desert floor, they can be seen from a long way away.

AGRICULTURAL AND IRRIGATION CANALS: can lead you to people.

SAND DUNES

It is less taxing to walk on the windward side of the tops of sand dunes than walking up and down them in an effort to stay on a straight course.

MOVEMENT

Travelling in the desert can be extremely hazardous. As a survivor you must consider the effect the environmental factors, your condition and the amount of food and water required will have on travel. *DO NOT* underestimate the climate or the terrain. In daytime, the scorching heat will make movement during impracticable. However, if you are travelling at night in rocky or mountainous deserts you may not see eroded drainages and canyons, which could result in you falling and sustaining serious injury.

Navigation At night use the stars and the moon to navigate by (see Signalling and Navigation Chapter). During the day you can use a compass or landmarks. However, be aware that in the desert, because of the glare and lack of landmarks, distances are difficult to estimate and objects difficult to size. Survivors should try to follow animal trails and hope they lead to rivers or watering holes. The wind can be used as a direction indicator: orientate yourself to any prevailing winds once you have established they are consistent and you know in what direction they are blowing.

Sandstorms can totally disorientate you. When the storm is over all the landmarks you were using may be obliterated or indistinguishable. You must mark

US MARINE CORPS TIPS

DESERT TRAVEL RULES

Travelling during the day in the desert can be a killer. US Marine Corps regulations are strict concerning movement in desert regions.

- Avoid the midday sun: travel only in the evening, at night or in the early morning.
- Do not walk aimlessly. Try to head for a coast, a road or path, a water source or an inhabited location. Try to follow trails.
- Avoid loose sand and rough terrain: they will cause fatigue.
- In sandstorms, lie on your side with your back to the wind, cover your face and sleep through the storm (don't worry – you won't get buried).
- Seek shelter on the leeward side of hills.
- Objects always appear closer than they really are in the desert. Therefore, multiply all your distance estimations by three.

your route before a storm so you can pick up the trail afterwards. Placing a stick to indicate direction will suffice.

Mirages can play havoc with your navigation. Be especially alert for mirages concealing objects, creating imaginary objects and for making red objects seem closer and larger. Remember, mirages are common during the heat of the day.

Clothing is extremely important in desert areas. You must have protection against sunburn, heat, sand and insects. *DO NOT* discard any clothing. Keep your head, legs and body covered at all times. (Diagram 94). Do not roll up your sleeves: keep them rolled down and loose at the cuffs to stay cool. Light-coloured flowing robes reduce high humidity between the body and the clothing; which helps to keep you cool

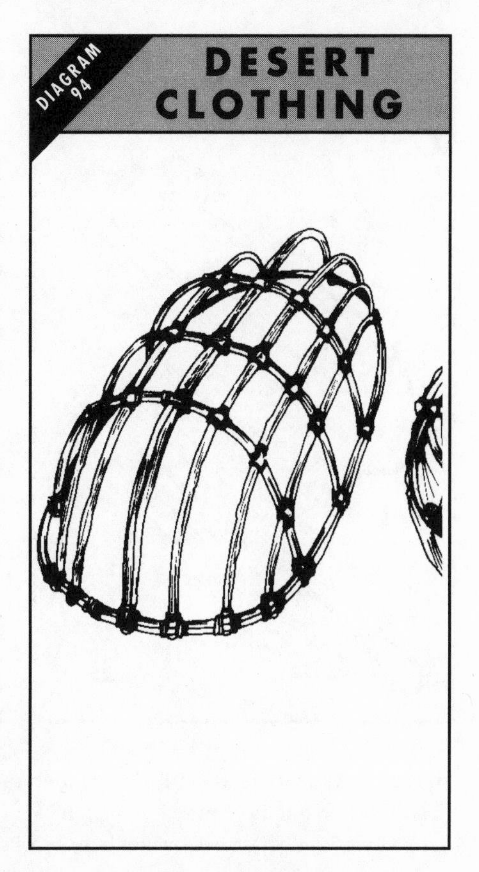

FRENCH FOREIGN LEGION

RULES FOR DESERT CLOTHING

The French Foreign Legion has over 100 years experience of desert fighting. Its men have learned the hard way how to dress for the desert.

- Keep your body well covered during the day.
- Wear long trousers and a long-sleeved shirt.
- Keep your head covered at all times.
- Wear a cloth neckpiece to protect the back of the neck from the sun.
- Wear clothing loosely.
- Open your clothing only in shaded areas.
- Take off your boots and socks only in the shade.
- Shake your boots before you put them back on in case a scorpion or spider has crawled into one of them.

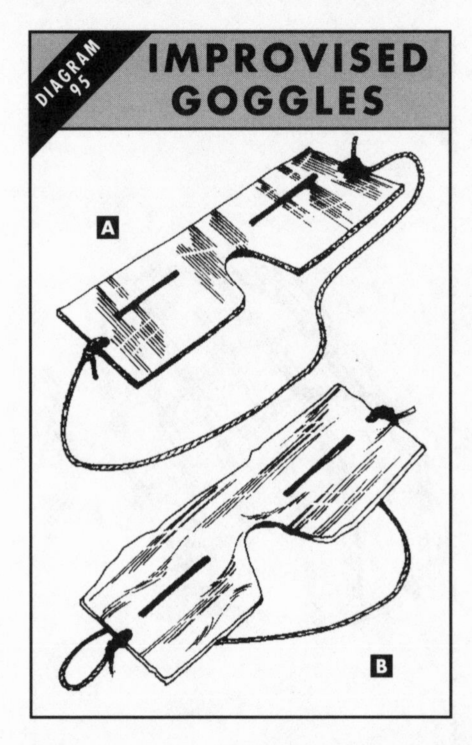

and limits perspiration. White clothing also also has the advantage of reflecting the sunlight.

Eye protection (Diagram 95) Wear sunglasses or goggles if you have them. If not, improvise a pair of sun shades, from material (A) or bark (B). Make the eye slits narrow. Reduce glare reflected from the skin by smearing soot from a fire below the eyes. *DO NOT* neglect eye protection. Sand and grit can blow into the eyes and cause injury.

FINDING WATER AND FOOD

It is vital you find water in the desert: without it you will die quickly. Without water you will last two and a half days at a temperature of 48 degrees C (120 degrees F) if you just rest in the shade, and up to 12 days if the temperature stays below 21 degrees C (70 degrees F). In a temperature of 48 degrees C (120 degrees F), you will be able to walk 8km (5 miles) without water before collapsing.

Relative to water, food is less important in the desert. Heat produces a loss of appetite. Do not force yourself to eat. As liquids are needed for digestion, try to eat moisture-containing foods, such as fresh fruit and vegetables, to maintain your body's water level.

Remember that food, especially meat, spoils very quickly in the heat of the desert. Be particularly alert for flies, which seemingly appear out of nowhere to settle on your uncovered food.

WATER

There are three things about water routes that you should bear in mind:

- Water always flows downhill.
- It grooves the face of the earth, making creek beds, canyons and washes.
- It encourages vegetation.

Water in the desert may be underground. Find a dry lake bed at it lowest part and dig into the ground with a spade, stick or rock. If you strike wet sand at once stop digging and allow the water to seep in. In dry river beds (Diagram

FRENCH FOREIGN LEGION

CONSERVING WATER

Water is the most precious commodity in the desert. French Foreign Legionnaires thus implement the following measures to reduce body fluid loss.

- Stay fully clothed: you will perspire a lot less.
- Do not use water for washing unless you have a regular supply.
- Do not rush around: you must keep perspiration down.
- Drink water in small sips not gulps. If water is critically low use it only to moisten your lips.
- Keep small pebbles in the mouth or chew grass to relieve thirst.
- Use salt only with water and only if you have a regular water supply.

96), find a bend and dig down on its lower side. If you don't have immediate success,stop digging and find another spot: conserve your energy.

Observe the terrain closely: the likeliest place to find water will be at the base of a hill or canyon. Greenery on canyon walls is an indicator of a seep in the rocks. Also look for vegetation, especially reeds, grass, willows, cottonwoods and palm trees: they usually mark permanent water sources.

DIAGRAM 96

DRY RIVER BED

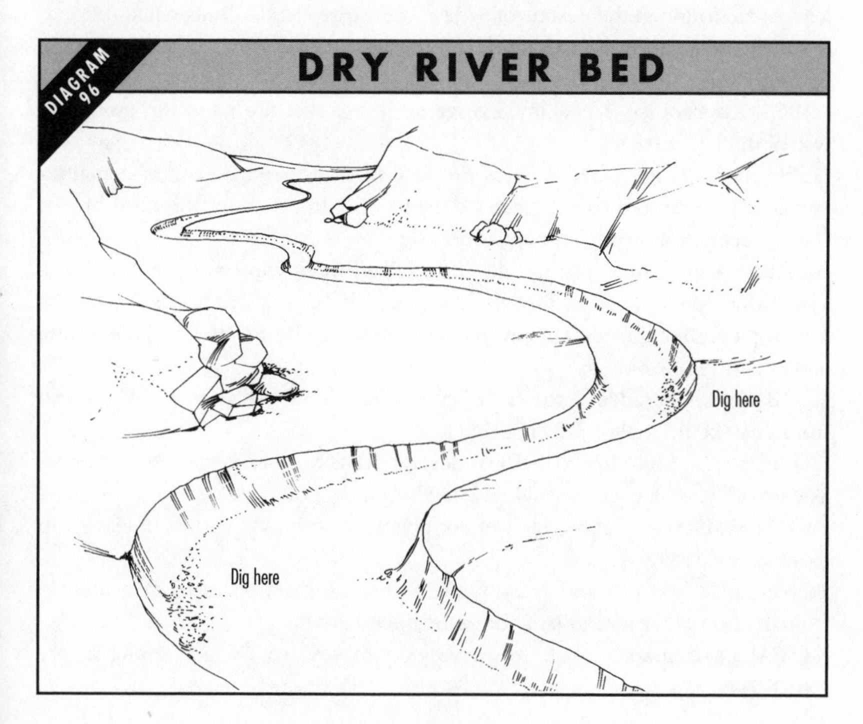

Desert plants can be valuable water sources in themselves. Peel off the tough outer bark of a cactus and chew on the liquid-filled inner tissue. The leaf stems of other desert plants, such as pigweed, contain water. Pigweed has fleshy reddish-green leaves and stems. In season it has yellow flowers and covers the ground in patches.

As well as finding water sources you can also make your own using stills and vegetation bags (see Water Chapter).

Remember to purify all water before you drink it (see Water Chapter), although any water obtained from plants, trees, shrubs, dew rain or snow will not need purifying.

FOOD

DO NOT EAT UNLESS YOU HAVE WATER. The US Air Force has a rule: if you can only get 0.5litres (one pint) of water a day you should not eat at all. You should not eat foods that contain proteins – which require water for digestion – unless you have sufficient supplies of water.

Plant food Availability of plant food varies according to geographical area. Date palms are found in most deserts and are cultivated by native peoples around oases and irrigation ditches. Fig trees (straggly trees with leathery evergreen leaves) are found in the deserts of Syria and Europe. Eat the fruits when they are ripe: when they are coloured green, red or black.

Learn to identify the following types of edible plants in the desert:

CAROB Appearance: has shiny, evergreen leaves that are paired in groups of two or three to a stem.

Edible parts: Its small red flowers produce leathery seed pods that contain a sweet, nutritious pulp which can be eaten raw. In addition, the hard brown seeds can be ground and cooked as porridge.

ACACIAS Appearance: thorny, medium-seized trees with very small leaflets. Their white, pink or yellow flowers form small globular flower heads.

Edible parts: use the roots for water; the seeds can be roasted and the young leaves and shoots boiled.

BAOBABS Appearance: large trees with huge, heavily swollen trunks. These trunks can be up to 9m (30ft) in diameter.

Edible parts: cut into the roots for water, the fruits and seeds can be eaten raw. The tender young leaves should be boiled.

DATE PALMS Appearance: tall, slender palms crowned with a tuft of leaves up to 4.8m (16ft) long

Edible parts: the fruits and growing tip of the palm can be eaten raw; the sap from the trunk is rich in sugar and can be boiled down.

MESCALS Appearance: thick, leathery, spiky leaves, from which spring a very long flower stalk.

Edible parts: the stalk is edible when cooked.
WILD GOURDS: the plant resembles a vine, with orange-sized fruits.
Edible parts: you can boil the unripe fruit to make it more edible. Cook young leaves, roast the seeds and chew the stems and shoots for their water.
CARRION FLOWERS Appearance: large plants with succulent stems that branch off into leaves like fat spines. They have star-shaped flowers covered in thick, shaggy hairs. The flowers give off a stench of rotting meat, hence their name.
Edible parts: you can tap the stems for water.
PRICKLY PEARS Appearance: have thick, pad-like leaves, yellow or red flowers and egg-shaped pulpy fruits.
Edible parts: the peeled fruits can be eaten raw, the pads must be cooked (cut away the spines). Roast the seeds for flour and tap the stems for water.

All desert grasses are edible. The best part is the whitish tender end that shows when the grass stalk is pulled from the ground; you can also eat grass seeds. *FLOWERS THAT HAVE MILKY OR COLOURED SAP ARE POISONOUS.*

Animal Food All desert animals – mammals, birds, reptiles and insects – are edible. You may have to lay traps for the larger animals and birds (see Food Chapter). However, keep a watch for owls, hawks, vultures and wolves, which often congregate around freshly killed animals. Chase them away and take the meat for yourself. If you are desperate, set a fire in heavy grass or sagebrush: after it has gone out you can look around for cooked rabbit or rat.

Rabbits and birds can be trapped, but rabbits can be smoked out of their holes by building a fire at the entrance. Be ready to club them when they come out. Snakes make a tasty meal, but remember they may be poisonous. You will often find them sunning themselves on rocks and ledges when the sun isn't too hot (when the sun is at its height they will stay in shaded areas). They are most active during the early morning or early evening.

US SPECIAL FORCES TIPS

EATING AND COLLECTING DESERT INSECTS

The US Green Berets know that insects are a valuable food source. Follow these guidelines and supplement your diet.

- Attract insects at night with a small light.
- Gather crawling insects by lifting up stones.
- The larvae of ants make good eating. Brush them from the undersides of stones into a container of water. The larvae will float to the top.
- Remove the wings and legs of grasshoppers and crickets before eating.
- Always cook grasshoppers before eating.

Lizards can also be eaten. Look out for them under flat stones at dawn, before the sun has warmed the air, and kill them by throwing stones at them or using a catapult (see Improvising Tools and Weapons Chapter).

Most people don't like the thought of eating insects, but they can make a tasty meal. However, avoid caterpillars as a few species are poisonous, and stay clear of centipedes and scorpions.

Desert cooking With a little imagination you can use the materials to hand and turn your raw meat into a tasty meal. Use the following cooking methods:

BOILING: make a boiling pot by scooping a hole in hard ground, lining it with leaves or other waterproof material, filling it with water, and then dropping red-hot coals from the fire into it. In mountainous deserts, you can find rocks with holes in them made by erosion. You can use them as boiling pots: preheat them on the fire before filling them with water. Large plant leaves can be bent into containers to hold water (they will not burn if held over a flame as long as you keep them filled with water). Barrel cacti can also be used as containers: mash the pulp inside them, scoop it out and use the shells for cooking.

FRYING PANS: they can be created from desert rocks. Find a flat and narrow rock, prop it up with other rocks and build a fire underneath it (make sure you wipe off all the grit and dirt before you heat it). The rock will heat up, allowing you to fry meat or eggs (raid birds' nests to get them) on it.

ROASTING: ensure you have a good bed of glowing coals, then place your meat on a green stick and hold it near the embers.

KEBABS: prepare a bed of hot, glowing embers. Cut your meat into even cubes and spear each one with a stick that has a sharpened end. Roast each meat cube evenly over the hot coals (make sure they do not slip off the sticks).

MUD BAKING: this is an excellent way of cooking small animals. Clean the carcass by removing the head, feet and tail. Leave the skin or feathers on the animal. Cover the carcass completely with mud or clay at least 2.5cm (1in) thick, place in a large fire and cover with coals. The animal may take up to an hour to cook, depending on its size. When the clay is hard and brick-like, remove it from the fire and break open the covering. When you do this the feathers or skin will break away at the same time, leaving a ready meal.

BAKING IN ASHES: a good method to cook an animal that will have its skin removed after cooking. Rake some ashes and embers to one side, place the food on top of the coals and then cover with the remaining coals.

GRILLING: dig a hole larger and wider than the animal to be cooked, approximately 0.3-0.9m (1-3ft) deep. Build a fire at the bottom of it and wait until it is hot. Then spread green poles over the top of the hole and place the meat on top of them. You can put small stones in the fire to radiate the heat.

EARTH OVEN: this method of cooking is used in the South Seas. Dig a hole 0.6m (2ft) wide and 0.6m (2ft) deep. Gather some wood and make a criss-cross pattern over the hole, laying one layer of sticks in one direction and one layer in another. Then lay a number of medium-sized stones on top of the sticks. Start a fire in the hole and let it burn until the stones turn white and fall into the hole. Arrange the stones in the hole and shovel out any pieces of burning wood, then cover the stones with a lot of green leaves that have been moistened with water and throw the food to be cooked on top. Cover the food with another batch of leaves and then cover the hole with earth to ensure no steam escapes. After about two hours the food will be cooked. This is an excellent method of cooking but requires speed when you are arranging the stones to ensure they do not cool.

DANGERS

The main threats to you in the desert will be the heat and dangerous animals and plants that can injure or poison you. However, with a little care you can cope with all these things.

Dangerous plants Most desert plants are protected by sharp thorns or spines. The spines have tiny hooks on them that will stick to your skin or clothing if you touch them. Give them a wide berth. Poison oak and poison ivy can cause intense skin irritations (see poisonous plants in Food Chapter). Wash skin thoroughly after exposure.

Insects Avoid ant nests (they can be identified as protruding mounds of soil). If you are bitten, a mud pack will soothe the rash and reduce the pain. Centipedes should also be avoided: their bites can be very painful and the effects can last up to two weeks. They hide under rocks during the day and move at night.

Scorpions stay under rocks during the day and move around at night, often into sleeping bags or boots. If you are camping use a tent with a sewn-in floor. Shake out your boots in the morning. There is no real field treatment for scorpion stings (the treatment for snake bites is ineffective – there is an immediate and overall reaction with scorpion stings). Fortunately, most adult victims of scorpion stings recover, though child fatalities are more common.

Spiders Spider bites cannot be treated effectively in the wild. You will just have to endure pain, nausea, dizziness and difficulty in breathing for a few days. It is far better to avoid spiders. Do not tease or try to capture them.

You may encounter the following spiders in desert regions:

BLACK WIDOW Appearance: small, dark and has a red, yellow or white marking on the abdomen.

Bite symptoms: severe pain, sweating, shivering and weakness. Can disable a victim for up to a week.

FIDDLEBACK Appearance: has a violin shape on the back of the head.

Bite symptoms: fever, chills, vomiting, joint pain and spotty skin. Tissue damage around the bite may lead to amputation in serious cases.
TARANTULA Appearance: large and hairy.
Bite symptoms: some pain, but the poison is fairly mild and not disabling.

Snakes Venomous snakes found in desert areas include the cobra, viper and rattlesnake. Your best protection against snake bites is to wear protective clothing – most bites are below the knee or on the hand or forearm. Do not put your hands into places you cannot see, do not try to catch a snake unless you are certain you can kill it, and always wear boots. Be careful where you tread. For treatment of snake bites see the First Aid Chapter.

Specific types of snake you may encounter in deserts include the following:
RUSSELL'S VIPER Appearance:tan or brownish-yellow coloured with spots ringed with black.
Size: 0.9-1.5m (3-5ft).
Temperament: bold, hisses loudly and strikes with great speed.
Distribution: west Pakistan, all of India, Thailand and southwest China.
SAND VIPER Appearance: yellow or pinkish colour.
Size: around 0.6m (2ft).
Temperament: because of their gliding gait, they can move across desert sands at great speed. Vicious if provoked.
Distribution: North Africa from Algeria to Egypt.
EASTERN DIAMONDBACK RATTLESNAKE Appearance: olive-green coloured with dark diamond-shaped markings with white or yellow edgings.
Size: 0.9-1.5m (3-5ft).
Temperament: bold, will defend itself by coiling, inflating its body with air and making a low hiss. In addition, its tail will usually vibrate.
Distribution: the eastern United States: North Carolina southward to Florida and west to Louisiana.
FER-DE-LANCE Appearance: usually brown or olive coloured, with faint hourglass markings along the back.
Size: 0.9-1.5m (3-5ft).
Temperament: it may coil its body before striking, but it can strike from any angle. Do not provoke.
Distribution: found throughout Central and South America.
WESTERN DIAMONDBACK RATTLESNAKE Appearance: light-brown coloured, with darker brown diamond-shaped markings. Its tail is cream to white and heavily marked with black rings.
Size: 0.9-1.5m (3-5ft).
Temperament: has a bold disposition, will readily defend itself.
Distribution: the United States, specifically Texas, Louisiana, Arkansas, southeast California. Also found in northern Mexico.

MOJAVE RATTLESNAKE Appearance: green or olive coloured.
Size: 0.6-0.9m (2-3ft).
Temperament: can strike without a warning, often giving a rattle afterwards.
Distribution: Mojave Desert in California and the Mexican highlands.
DEATH ADDER Appearance: light-brown or reddish coloured with darker crossbands. The tail is yellow and short.
Size: 0.35-0.9m (1.5-3ft).
Temperament: bad tempered, will bite quickly with little provocation.
Distribution: throughout the Far East and also found in Australia.
EGYPTIAN COBRA Appearance: can be black, brown or yellow coloured. Has darker crossbands.
Size: 1.5-2.4m (5-8ft).
Temperament: aggressive.
Distribution: most of North Africa and the southwestern Arabian Peninsula.
SAW-SCALED VIPER Appearance: rough-scaled and ranges in colour from pale red to sandy-brown, with darker markings and white blotches
Size: 0.4-0.55m (1.3-2ft).
Temperament: extremely vicious.
Distribution: North Africa and India.
AUSTRALIAN BROWN SNAKE Appearance: yellowish-grey to brown, with a pale belly.
Size: 1.5-2m (5-6ft).
Temperament: aggressive.
Distribution: Australia and New Guinea.
TIGER SNAKE Appearance: large-headed, thick-bodied, has greenish-yellow, grey or orange-brown bands.
Size: 1.3-1.6m (4-5.5ft).
Temperament: aggressive.
Distribution: Australia and Tasmania.

Lizards Both the gila monster and beaded lizard (both around 45cm (18in) long) are poisonous. The gila monster has a large rounded head, thick chunky body, short stumpy tail and is brightly patterned yellow. The beaded lizard is darker and larger with a slender tail. Both these creatures are docile and will run away from you. *DO NOT* tease or corner them, their bite is very poisonous. If bitten treat as for snake bite.

Animal bites Mammals can carry rabies. If you are bitten, immediately scrub the bite area with soap and water and apply disinfectant (if you have it). If a member of your party has rabies and is in the advanced stage, isolate him or her and tie him or her down. The person will unfortunately certainly die – do not touch the body after death.

Animals in the advanced stage of rabies, especially dogs, will be violent, will stagger and foam at the mouth. If you are the victim of an unprovoked attack, you have good reason to suspect that the animal has rabies. Hospital treatment for rabies must be started within one or two days to be effective.

Dehydration Be careful your body does not dehydrate. In very high temperatures and low humidity sweating may not be noticeable because it evaporates quickly. You must try to retain sweat on the skin to improve the cooling process – avoid direct sunlight on the skin. *STAY CLOTHED.* Remember, thirst is not always an adequate warning of dehydration. Keep up your fluid intake.

Heat injuries Result from a deficiency of salt or water during heavy sweating, apart from heatstroke, which results from a failure of the sweating mechanisms.

Heat cramps are caused by excessive loss of salt from the body when you have been sweating heavily. Cramps are painful spasms of the muscles, usually legs, arms and abdominals. They can be mild or severe. Treat by drinking large amounts of water.

Heat exhaustion is caused by excessive loss of water and salt from the body through sweating. The skin becomes cold and wet with sweat, with accompanying headache, dizziness, weakness and loss of appetite. Can be fatal if untreated. Place victim in the shade, massage and elevate legs to return blood to the heart. Give large quantities of water to drink.

Heatstroke occurs when the body loses its ability to cool itself by sweating. The skin becomes hot and dry. Victim may suddenly collapse or experience headache, dizziness or even delirium before becoming unconscious. Heatstroke is potentially fatal. Treatment is aimed at lowering the body temperature as

US MARINE CORPS TIPS

TREATMENT FOR ALL HEAT INJURIES

Desert heat injuries can be potentially fatal. You must administer treatment immediately. Follow these US Marine Corps guidelines:

- Your primary aim is to lower patient's body temperature immediately.
- Place patient on his or her back in a shady place.
- Loosen patient's clothing.
- Sprinkle patient with water.
- Fan the patient.
- If patient is conscious and rational, give him or her a salt tablet and plenty of cool (not cold) water to drink.
- Do not administer any stimulants.

quickly as possible. Place patient in shade, remove clothing and sprinkle body with water from head to foot. Fan to increase the cooling effect. Massage the legs and arms to stimulate circulation.

In the desert you are vulnerable to a host of insect-borne diseases, such as malaria, sandfly fever, typhus and plague. You must try to employ preventative medicine measures and adequate personal hygiene and sanitation. Guard against cuts and scratches: in the desert they can become infected very easily. To prevent intestinal diseases, clean all cooking and eating utensils, dispose of garbage and human wastes and protect food and utensils from flies.

Do not expose your flesh to the elements or to flies; try to wash your feet and body daily; and change your socks regularly. You must check yourself for signs of any injury, no matter how slight. Remember dust and insects can cause infection of minor cuts and scratches.

For sanitary reasons you must bury all garbage and human wastes, but remember you must bury them deep because shallow holes can become exposed in areas of shifting sands.

BUILDING SHELTERS

Shelter is extremely important in the desert, both to protect you from heat during the day and to keep you warm during the intense cold of the night. Try to copy the natives when it comes to shelters: light shelters that have the sides rolled up to take advantage of any breeze. You might not be able to build an exotic tent but you can make use of the same principles the locals employ

US ARMY TIPS

SIMPLE DESERT SHELTER

If you have a piece of canvas, a poncho or a parachute-like cloth, you can erect two simple desert shelters. As well as a mound or an outcrop of rock, you will need at least two sticks to support the extended end of the canvas

Rock method

- Find an outcropped rock.
- .Anchor one end of your material on the edge of the outcrop with rocks.
- Extend and anchor the other end to give you the best available shade.

Mound method

- Construct a mound of sand or use the side of a sand dune for one side of the shelter.
- Anchor one end of the material on top of the mound with rocks or weights.
- Extend and anchor the other end of the material to give you shade.

FRENCH FOREIGN LEGION

BUILDING DESERT SHELTERS

Knowing when and where to build desert shelters can save you a lot of time and energy. Follow the advice of the French Foreign Legion.

- Build shelters during the early morning, late evening or at night. It is less physically taxing.
- Try to build a shelter near fuel and water if possible.
- Do not construct shelter at the base of steep slopes or in areas where you risk floods, rock falls or battering by winds.
- Build shelters away from rocks which store up heat during the day (you may wish to move to rocky areas during the night to take advantage of the warmth).

Natural shelters Can be scarce in the desert, and limited to the shade of cliffs and the lee sides of hills, dunes or rock formations. Caves are a good shelter in rocky areas, but you will have to look carefully for them because they are small and easy to miss. Look out for crevasses or jutting ledges. Caves are cool and they may contain water. However, they can also contain animals: rats, mice, snakes and rabbits. Although these are all food sources and some individuals may be attracted to the idea of having a food store at their fingertips, sensible survivors will be aware of the dangers of bites and stings. Therefore, stay near the entrance.

DIAGRAM 97

EMERGENCY SHELTER

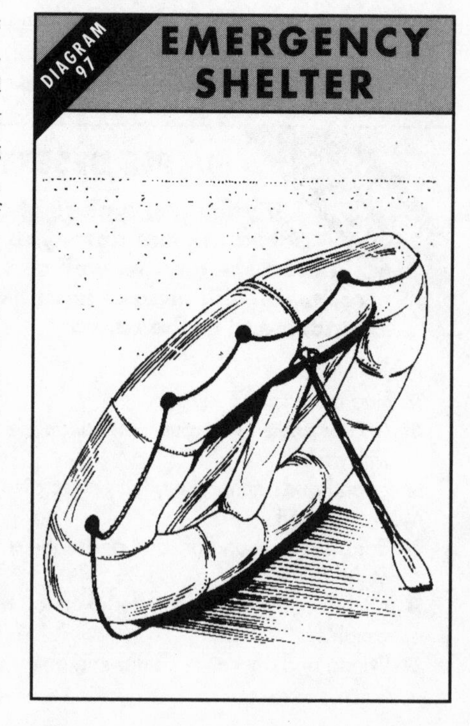

In flat, open deserts natural shelters are hard to come by. However, gather together tumble weeds and mat them together. Use any vegetation that you can find to make into a shelter.

In some deserts, the sand 0.5m (2ft) below the surface can be as much as 20 or 30 degrees cooler than the air above it. Dig into the sand and construct a trench 1m (3ft) deep dug in a north-south direction to provide shade during

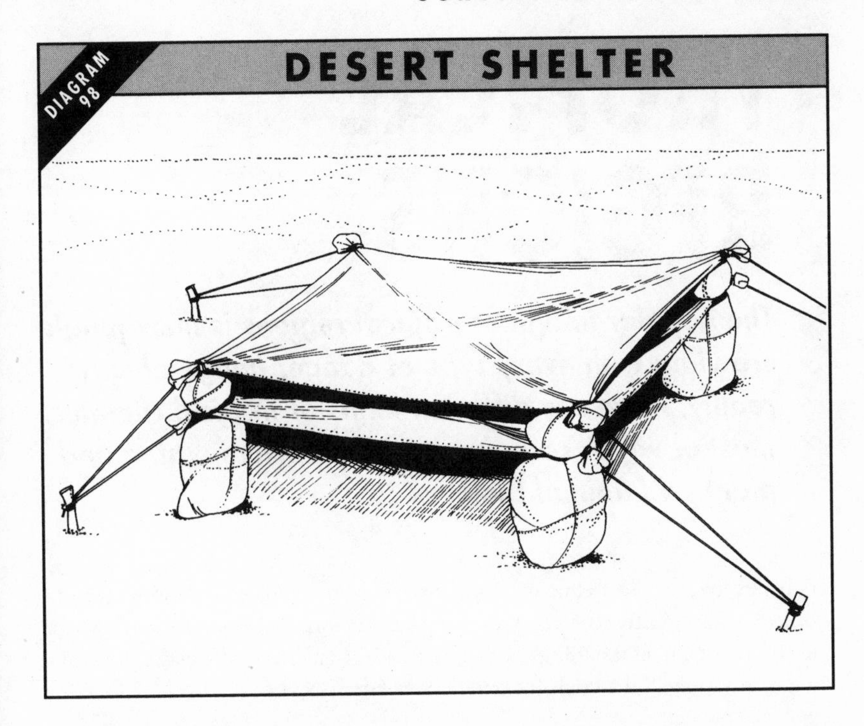

the day. Cover it to provide more protection. When building your shelter you must bear in mind three points:

- ☐ Keep an eye on the weather. If a storm is threatening, avoid gullies, washes or areas with little vegetation. They are prone to floods and high winds.
- ☐ Poisonous snakes, centipedes and scorpions may be hiding in brush or under rocks.
- ☐ Do not make camp at the base of steep slopes or in areas where you run the risk of floods, rockfalls or heavy winds.

Simple shelters If you are the survivor of an air crash you may be able to salvage aircraft parts and life rafts to make a shelter. For example, an inflatable raft can be tilted to give protection against the sun (Diagram 97).

Try to build a shelter that has more than one layer so the resulting airspace reduces the inside temperature of the shelter (Diagram 98). You should place the floor of the shelter about 46cm (18in) above or below the desert surface to increase the cooling effect. Try to use a white material as the outer layer of the shelter, and the sides of the shelter should be movable to protect you during cold and windy periods and to allow ventilation in the extreme heat.

One more thing: build your shelter to take advantage of a breeze – it will keep you cool and keep insects away.

TROPICAL REGIONS

The popular image of tropical regions is thick jungle crawling with every type of danger imaginable. In reality, there are different kinds of tropical climates, most of which contain an abundance of water, and plant and animal food.

There are five tropical climates: rain forests, semi-evergreen seasonal forests, tropical scrub and thorn forests and tropical savannas. The jungle can provide the survivor with all the things he or she needs to maintain life: water, food and an ample supply of materials for building shelters.

CHARACTERISTICS OF THE TERRAIN

Rain forests A term applicable to the jungles of South America, Asia and Africa. Evergreen trees predominate and grow to the height of 45-54m (150-180ft). The bases of many trees have ridge buttresses and their leaves are leathery and dark-green, while their flowers are inconspicuous and coloured green or white. On river banks or in clearings, where sunlight reaches the ground, there is dense growth which is often impenetrable (so-called secondary jungle). However, where the jungle canopy is thick there is little undergrowth to impede movement. There is also an abundance of climbing plants, which usually hang like cables or loops.

There is high rainfall – around 203cm (80in) a year – evenly distributed throughout the year. Most plants that grow in the rain forests are woody and are comparable in size to trees. For example, the air plants and vines that grow on the trunks and branches of trees are tree-like. Grasses and herbs are rare, and any plants on the forest floor usually consists of woody plants: sapling and seedling trees, shrubs and young woody climbers.

Bamboos, which are really grasses, grow to large proportions in the rain forests: as high as 24m (80ft) in some cases, with thick stalks that can contain water (see below). They usually grow in dense thickets that are almost impossible to penetrate.

BRITISH SAS TIPS

GENERAL CHARACTERISTICS OF TROPICAL REGIONS

Get to know the nature of the terrain you are in. Tropical regions have a number of common characteristics. Here's a quick check list.

- High temperatures and oppressive humidity.
- Heavy rainfall, often accompanied by thunder and lightning, which causes rivers to rise rapidly and turns them into raging torrents.
- Hurricanes, cyclones and typhoons develop over sea areas and rush inland, resulting in tidal waves and devastation.
- There is a 'dry' season (during which it rains only once a day) and a monsoon season (when it can rain for days or weeks continuously).
- Tropical day and night are of equal length.

There is no winter or spring in the rain forest: this means that the vegetation looks the same at any time of the year. Around the fringes of the forest, where there are clearings and areas of abandoned dwellings, there are many edible plants. However, in the middle of the virgin forest, where the trees are so tall and the fruits and nuts are out of reach, finding food is more difficult.

Semi-evergreen seasonal forest The semi-evergreen seasonal forests in Central and South America and Africa are essentially similar to the monsoon forests of Asia. There are two strata of trees, an upper storey 18-24m (60-80ft) high and a lower storey 6-13.5m (20-45ft) high. There is a seasonal drought which causes leaf fall, and a monsoon season.

The monsoons of India, Burma and Southeast Asia are of two types: the dry monsoon and the wet monsoon. The dry monsoon takes place from November to April, when the northern winds from central Asia bring long periods of clear weather and sparse rainfall. The wet monsoon takes place between May and October, when the southern winds from the Bay of Bengal bring heavy rain that lasts for days or weeks at a time, which causes the foliage to reappear overnight.

Tropical scrub and thorn forests There is a definite dry season which causes the leaves to fall off the trees, and rains appear mainly as downpours from thunderstorms. The average height of trees is 6-9m (20-30ft), and plants with thorns predominate. This type of forest is restricted to parts of Central and South America, southern Africa, India and northern Australia.

Tropical Savanna Lie wholly within the tropical zone in South America and Africa. The savanna looks like a broad, grassy meadow with trees spaced at wide intervals. The grasses can be quite tall, and form bunches, with spaces between each grass plant. The soil can be frequently red and scattered trees look stunted. There is a wet season in the first half of the year and a dry season in the second half.

Swamps Saltwater swamps occur when coastal areas are subjected to tidal flooding. They are filled with mangrove trees, which have tangled roots, above and below the waterline. Tidal rise may be as much as 12m (40ft) – determine the high-tide level by the line of salt and debris on trees.

Freshwater swamps can be found in low-lying inland areas. Often dotted with islands, they are a mass of thorny undergrowth, reeds, grasses and palms.

MOVEMENT

Before considering travel, you should consider your chances of being found and rescued from your present location: you don't want to travel if you don't have to. Food and water should not be a problem where you are: they usually abound in the jungle. Moving through the jungle can be slow and exhausting. Are you in the right physical shape to tackle the journey?

In many cases rivers, trails and ridge lines are the easiest routes to follow, though there are some problems associated with them. Rivers and streams can be overgrown, making them difficult to reach and impossible to raft. The waterways themselves may be infested with leeches and dangerous fish and reptiles. Trails can have traps or animal pits on them, and they can also lead to a dead

US SPECIAL FORCES TIPS

MOVING THROUGH THE JUNGLE

US Green Berets move swiftly and stealthily through the jungle. Follow their advice: do not make things unnecessarily difficult for yourself.

- Avoid thickets and swamps; move slowly and steadily through dense vegetation.
- Only move through the jungle in daylight.
- Use a stick to part vegetation to reduce the possibility of disturbing ant or scorpion nests with your hands or feet.
- Do not grab brush or vines to help you up slopes or over obstacles: their thorns and spines will cause irritation and they may not hold your weight!
- Do not climb over logs if you can walk round them: you may slip and get injured or step on a snake.
- If using a trail, watch for disturbed areas – it may be a trap or pitfall.
- Do not follow a trail that has a rope barrier or grass net across it: it may lead to an animal trap.

> **DANGER**
> **QUICKSAND!**
>
> *Usually located near the mouths of large rivers and on flat shores, you can easily get caught in quicksand.*
> *If you do, adopt a spread-eagled position to help disperse your body weight and stop you sinking. Spread out and swim or pull along the surface.*
> *DO NOT PANIC: you will sink quicker if you do.*

end or into swamps or thick bush. More hazardous, the vegetation along a ridge may conceal crevices or even extend out past cliff edges.

Despite the hazards involved, waterways often offer the least troublesome route of travel. If you can, find a stream and travel downstream to a larger body of water. Though following a stream may mean fording water and cutting through dense vegetation, a stream gives you a definite course that will probably lead to some sort of habitation. It is also a source of food and water and may allow you to travel by raft.

Crossing a stream You must study the stream to find a place to ford it safely, and look at the opposite bank and ensure it can be climbed. When selecting a fording site you should look out for a travel course that leads across the current at a 45-degree angle downstream (Diagram 99). *NEVER* attempt to ford a stream directly above, or close to, a deep or rapid waterfall or a deep channel. The stream should be crossed where the opposite side is comprised of shallow banks or sand beds. Avoid rocky places if you can, as a fall could injure you (though the odd rock that breaks the current can be helpful): a broken ankle or leg. Remember, deep water need not be a bad thing: it can run slowly and be safer than shallow water.

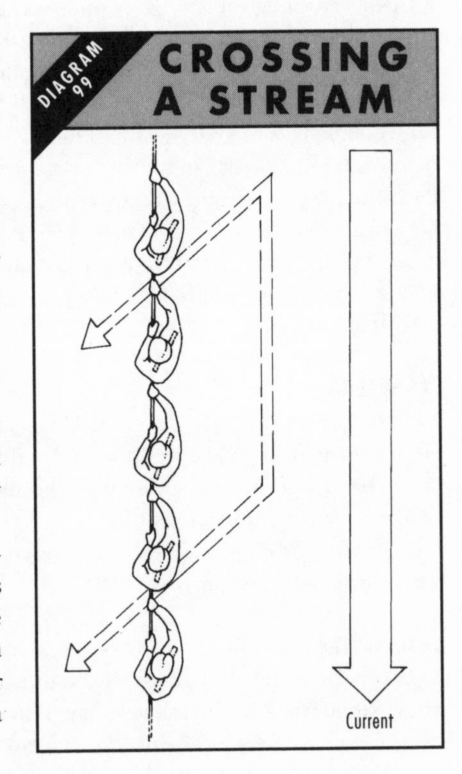

If the stream appears treacherous, use the method of fording as shown in Diagram 99 . If you are on you own, use a pole to give you greater balance (drag it in the water on the upstream side), or seek a safer place to cross.

US ARMY TIPS

TOOLS FOR JUNGLE TRAVEL

Movement through the jungle is much easier if you have one or more of the following items of equipment recommended by the US Army.

- A machete to cut through vegetation, collect food and cut logs to make a raft.
- A compass to maintain your direction.
- Medicines to treat fever and infection.
- Boots or stout shoes to make walking easier and to protect your feet.
- A hammock to reduce the time needed to prepare a bed above the jungle floor.
- Mosquito netting to provide protection against insects.

Machete One of the best aids to survival in the jungle. When using it, cut at a down and out angle, not flat and level: it requires less effort.

Ridges Travelling on a ridge is often easier than in a valley: a ridge has less vegetation, fewer streams and swamps. It allows you to survey the terrain and pick out landmarks. Ridges usually have game trails along their tops.

Camp sites Because darkness falls quickly in the jungle, you must set up camp before sunset. Do not camp too near a stream or pond during the rainy season: flash flooding can erupt without warning. Stay away from dead trees or trees with dead limbs which might fall on you. Be sure to cut away all the underbrush around your camp site to give you room and to allow your fire to ventilate. This will also reduce insects, hiding places for snakes and make you more visible to air rescue aircraft.

DANGERS

Most people imagine that the main dangers in the jungle come from snakes and large wild animals. While it is true that both are present in tropical jungle areas, the main danger to the survivor in the tropics is from insects that transmit diseases or have poisonous bites or stings. Your main adversaries are ticks, mosquitoes, fleas, mites, leeches, spiders, scorpions, centipedes, chiggers, wasps, wild bees and ants.

Insects Ticks and fleas are blood-sucking parasites that can carry infectious diseases. Ticks thrive in grassy areas. Brush them off your clothing and check your skin for them at least once a day. If they have attached themselves to your skin, apply treatment as described in the First Aid Chapter.

US ARMY TIPS

PREVENTING BITES AND STINGS

Insects are perhaps the greatest threat to the health of the survivor in tropical regions. Take measures to counter them.

- Use insect repellent, if you have it, on all exposed areas of the skin and on all clothing openings.
- Wear clothing all the time, especially at night.
- Cover your arms and legs. Wear gloves and a mosquito head net if possible to give you extra protection.
- Camp well away from swamps.
- Sleep under mosquito netting if you have it. If not, smear mud on your face to keep the insects away.

Fleas are found in dry, dusty shelters. They will burrow under your toenails or skin to lay their eggs. Remove them with a sterilised knife. Rats carry fleas that transmit plague, and rats can also cause jaundice and other fevers.

Red mites carry typhus fever. They burrow into the ground, and are especially common in tall grass and stream banks. Do not lie or sit on the ground. Clear your camping site and burn it off.

Centipedes, scorpions and spiders thrive in the jungle. Scorpions can be found beneath stoves and the loose bark of dead trees, but they can also get into shoes left on the ground in the night. Always shake out your shoes, socks, and clothing before putting them on.

Chiggers, wasps, wild bees and ants can harm you. Biting ants live in the branches and foliage of tropical trees, especially the hanging plants attached to mangrove branches. Be careful not to camp near an ant hill or ant trail. *ALWAYS KEEP YOUR BOOTS ON*: your footwear acts as protection against mites, ticks and ants.

Snakes There are many types of venomous snakes in tropical areas, including vipers, cobras, tropical rattlesnakes, mambas and kraits. Some cobras can spit poison as well as bite. If the poison gets into your eyes or an open cut, wash out wound immediately with water or even in an emergency with urine. Snakes will not usually bother you, but do not handle or provoke any snake and treat all snakes as poisonous. Note: a few tropical snakes, such as the bushmaster and mamba, will attack for no apparent reason.

The main types of dangerous snakes you are most likely to encounter in tropical regions are listed below:

GABOON VIPER Appearance: geometric patterns of blacks, tans and blues.
Size: 1.2-1.5m (4-5ft).

BRITISH SAS TIPS

SNAKES AND YOUR SAFETY

SAS soldiers are skilled jungle operatives. They treat the jungle and its inhabitants with respect, and know how to deal with the snake threat.

- Be careful where you step: snakes are often sluggish and can be stepped on.
- As some snakes live in trees, be careful when you pick fruit or part bushes.
- Do not provoke, corner or pick up a snake.
- Use a stick to turn over stones, not your hands.
- Wear stout boots if you have them. Many snake fangs cannot penetrate boot leather.
- Always check bedding, clothes and packs before putting them on. Snakes can crawl inside them.
- If you encounter a snake, *STAY CALM* and back off. In most cases the snake will want to escape.
- To kill a snake, use a long stick and strike it on the back of the head. Make sure you finish it off: a wounded snake is ferocious.

Temperament: will coil and strike quickly when approached.
Distribution: tropical rain forests of Sierra Leone, Sudan, Angola and Natal.
PUFF ADDER Appearance: light or dark-brown coloured, with white or yellow chevrons.
Size: 0.9-1.2m (3-4ft).
Temperament: strikes with lightning speed.
Distribution: most of Africa, but not the hot and arid regions of the desert.
RHINOCEROS VIPER Appearance: large, heavy bodied and coloured with many hues of pinks, blues and greens.
Size: 0.6-1.2m (2-4ft).
Temperament: strikes with lightning sped when approached.
Distribution: the tropical rain forests of Liberia, Uganda and Zaire.
BUSHMASTER Appearance: dark-brown or tan coloured, with some pink hues and black blotches along its back.
Size: 1.8-2.1m (6-7ft).
Temperament: may remain motionless until touched, will attack viciously if cornered. May sometimes attack for no reason.
Distribution: found throughout Latin America, mostly in forests at low altitudes.
COTTONMOUTH Appearance: young are coloured with bands of copper, light-brown and dark-brown. In adults, the bands may be faint to total black.
Size: 0.9-1.2m (3-4ft).
Temperament: fierce, do not annoy.

Distribution: found throughout the southern United States in swampy areas, lakes, streams and rivers.

GREEN TREE PIT VIPER Appearance: brilliant green coloured.

Size: 0.6-0.9m (2-3ft).

Temperament: not aggressive, but because they live in shrubs and trees bites are common (they are difficult to see).

Distribution: common throughout Southeast Asia.

JUMPING PIT VIPER Appearance: brown and black coloured, short body.

Size: 0.6-0.9m (2-3ft).

Temperament: will strike readily, often with such force that its body will leave the ground.

Distribution: throughout Latin America and southern Mexico.

MALAYAN PIT VIPER Appearance: reddish-brown back, dark-brown crossbands and pinkish-brown on the sides.

Size: 0.6-0.9m (2-3ft).

Temperament: calm disposition, but will bite if stepped on. It bites many people each year, largely because it lives in populated areas, e.g. on rubber plantations.

Distribution: common throughout Southeast Asia.

TROPICAL RATTLESNAKE Appearance: dark-brown with diamond-shaped markings down its back and dark stripes along its neck.

Size: 1.2-1.5m (4-5ft).

Temperament: will strike readily, coiling and elevating its head high above the coil. It may not rattle its tail before doing so.

Distribution: southern Mexico and all of Latin America except Chile.

WAGLER'S PIT VIPER Appearance: green coloured with black-edged scales. Has a stout body.

Size: 0.6-0.9m (2-3ft).

Temperament: placid.

Distribution: Thailand, Borneo, Indonesia, Malaysia and the Philippines.

KRAIT Appearance: coloured bright greyish to black, with narrow white crossbands and a white belly.

Size: 0.9-1.8m (3-6ft).

Temperament: not aggressive, but its venom is deadly.

Distribution: Asia and Southeast Asia.

CORAL SNAKES Appearance: coloured with vivid blacks, reds and yellows. Has a small head.

Size: 0.3-0.9m (1-3ft).

Temperament: placid, will not bite unless stepped on or picked up.

Distribution: southern United States and Latin America.

KING COBRA Appearance: olive- or light-brown coloured, large cobras can stand 0.9-1.2m (3-4ft) off the ground. Extends hood when it does so.

Size: 2.1-2.7m (7-9ft).

Temperament: aggressive, especially when guarding eggs.
Distribution: throughout Asia and Southeast Asia.
MAMBAS Appearance: green or dark-grey coloured, small head, slender body.
Size: 1.5-2.1m (5-7ft).
Temperament: quick to strike; have been known to attack without provocation.
Distribution: Africa south of the Sahara.
BOOMSLANG Appearance: green, brown or black coloured, very slender. It inflates its throat when alarmed.
Size: 1.3-1.5m (4-5ft).
Temperament: aggressive.
Distribution: Africa south of the Sahara.
TAIPAN Appearance: light- or dark-brown coloured, with yellowish-brown sides and belly.
Size: 3.5m (11ft).
Temperament: ferocious.
Distribution: open and forested parts of northern Australia.

Pigs All tropical areas contain wild pigs. These pigs have an aggressive nature and are omnivores. They will eat any small animals they can catch, though they feed mainly on roots, tubers and other vegetable foods. The main kinds of pigs in the tropics are the New World peccary, the Indian wild boar, the Babirussa of Celebes and the Central African Forest Hog. The peccary variety is divided into two main species: the 'white-lipped' peccary and the 'collared' peccary. Both kinds are coloured black, but the more ferocious 'white-lipped' peccary is black, with white under the snout. The 'collared' peccary has a grey or white band around the body where the neck joins the shoulder.

Though they are small – around 38cm (15in) high – wild pigs should be treated with caution. The 'collared' peccary often travels in groups of 5-15, and in these numbers they can easily deal with a jaguar, cougar or human. It is best to try to kill them with a spear trap (see Food Chapter). Do not try to tackle them yourself: their tusks can inflict severe injuries to the legs, which can be dangerously close to the femoral artery on the upper leg.

When preparing them for eating, note that both types of peccaries have musk glands that are located 10cm (4in) up from the tail on the spine. This gland must be removed soon after the animal has been killed, otherwise the flesh will become tainted and unfit for consumption.

Crocodiles and alligators They often lie on banks or float like logs with just their eyes above the water. Be careful when fording deep streams, bathing or getting near water. Avoid these animals at all times: their tails can inflict a scything blow and their jaws can crush you. If you have to get into the water, move slowly. Thrashing about will attract them.

Leeches Leeches are mainly aquatic and are found in freshwater lakes, ponds and water holes. They are attracted by disturbances in the water. Land leeches have a large appetite for blood and are easily aroused by a combination of colour, odour, light and temperature sense. Some leeches living in springs and wells may enter your mouth or nostrils when you are drinking and cause bleeding and obstruction.

Their bite is not painful and they will drop off you when they have had their fill of blood. However, if you are covered in them you must take immediate action. You can remove leeches with dabs of salt, alcohol, the burning end of a cigarette, an ember or a flame. *DO NOT PULL THEM OFF: IF YOU DO YOU MAY PULL THE HEAD OFF AND LEAVE THE JAWS IN THE BITE, WHICH MAY THEN QUICKLY TURN SEPTIC.*

Fish Be careful in the dry season when water levels are low. South American rivers can contain piranhas (up to 50cm (20in) long, deep-bodied, thickset, with razor-sharp teeth). Sharks have been known to attack humans in saltwater estuaries, bays or lagoons. Barracuda (see Survival at Sea Chapter) have also been known to attack in murky or clouded waters.

Candiru This is a minute Amazonian fish, about 2.5cm (1in) long, slender and almost transparent. It can swim up the urethra of a person urinating in the water. It then gets stuck by its dorsal spine. Though the chances of this happening are remote, you should not urinate in the water.

Reefs and shores Do not walk barefoot on coral reefs, which can cut your feet to ribbons. Fine needles from sponges and sea urchins may get into your skin

US ARMY TIPS

REEF DANGERS

Tropical reefs can be very hazardous places for survivors. Learn to recognise the following dangers associated with reefs:

- Avoid cone shells and terebra snails, which live under rocks, in crevices of coral reefs and along rocky shores and protected bays: they are poisonous.
- Do not use your hands to gather large clams. They will hold you if they clamp down on your fingers.
- When crossing deep portions of a reef, check the reef edge for sharks, barracudas and moray eels. Moray eels latter hide in dark holes. They are vicious and aggressive when disturbed.
- Keep clear of all tropical reef octopuses: they are poisonous.

and fester. You may also tread on a stonefish (see Survival at Sea Chapter), which will cause agonising pain, even death. .

Always use a stick to probe dark holes. Slide your feet along the bottom of muddy or sandy bottoms of rivers and seashores to avoid stepping on stingrays or other sharp-spiked animals.

Diseases The following diseases are prevalent in the tropics. They can all cause you harm and some can be fatal. Prevention is better than cure: get immunised before you travel if you can.

Bilharzia is a disease of the bowel or bladder, transmitted by microscopic fluke or worm. It enters the body through infected drinking water or broken skin. The main symptom is an irritation of the urinary tract, and it can be treated with doses of *NIRIDAZOLE.*

Hookworms are larvae that penetrate the skin, usually through the feet, or enter the body via infected drinking water. They cause lethargy and anaemia. The drugs *ALCAPAR* and *MINTAZOL* will cure the condition as will a decoction of bracken.

Amoebic dysentery is transmitted via contaminated water and uncooked food. Sufferers will feel listless and fatigued; their faeces will be solid but foul-smelling and contain blood and mucus-like red jelly. To treat, administer fluids, rest and take *FLAGYL.*

Malaria is an insect-borne disease which produces fever. Can be treated with *QUININE, PALUDRINE* and *DARAPRYN.*

Dengue fever is not treatable with drugs. However, one attack by this insect-borne disease will usually result in immunity. You will suffer headaches, pains in the joints and a rash for about a week.

Yellow fever produces vomiting, pains, fever and constipation. The treatment for this insect-borne disease is rest and nursing.

Typhus is transmitted by body lice or rat fleas. The disease results in vomiting, headache, nausea, a body rash, delirium, coma and death. Treatment includes antibiotics.

FINDING WATER AND FOOD

Both water and food are plentiful in tropical regions. Water for example is available from streams, springs lakes pools and swamps, though you must remember to purify it (see Water Chapter). There are many varieties of fruit and vegetable in the jungle and you can also trap the large variety of animal life that lives in tropical areas.

WATER

Surface water is available in the tropics in the form of streams, ponds rivers and swamps. In savannas during the dry season you may have to dig for water (see

AUSTRALIAN SAS TIPS

ENSURING WATER FROM VINES IS SAFE TO DRINK

Australian SAS soldiers are experts in jungle warfare. Use their simple guide to determine whether vine fluids are safe to drink.

- Nick the vine and watch the sap run from the cut.
- If the sap is milky, discard the vine.
- If the sap is not milky, cut out a section of the vine, hold it vertically and watch the liquid as it flows out.
- If it is clear and colourless it may be drinkable; if it is milky it is not.
- Let some of the liquid flow into the palm of your hand and observe it.
- If it doesn't change colour taste it.
- If it tastes like water or is sweet or woody, it should be safe to drink.
- Liquid with a sour or bitter taste should be avoided.

water section in Surviving in the Desert Chapter). You must purify and filter this water. Many plants have hollow portions which collect water: keep a look out for hollow sections of stems or leaves, Y-shaped plants (palms or air plants) and cracks and hollows. This water must also be purified.

Some varieties of vines are water sources, and water can be drunk from them without purification. When you are drinking from a vine, do not touch the bark with your mouth as it may contain irritants.

Bamboo (Diagram 100) Green bamboo often contains trapped water. Shake the bamboo: if a sloshing sound is heard it contains water. Cut off the end of a section that has water in it and drink or pour from the open end (A), though before you do look at the inside of the bamboo that contains the water. If the water is clean and white you can drink it; if it is brown or black or has any discolouration or fungus you must purify it before drinking.

You can also collect water by cutting the top off green bamboo then bending it and staking it to the ground. Place a container under it to catch the water (B).

Banana plants (Diagram 101) These plants contain drinkable water (A). Make a banana well out of the plant stump by cutting out and removing the inner section of the stump (B). Place a leaf from the banana plant over the bowl while it is filling: this prevents contamination by insects.

Coconuts Coconuts contain a refreshing liquid (this milky substance is safe to drink). The best coconuts to use are green, unripe over about the size of a grapefruit. The fluid can be drunk in large quantities without harmful effects,

DIAGRAM 100
WATER FROM BAMBOO
A
B

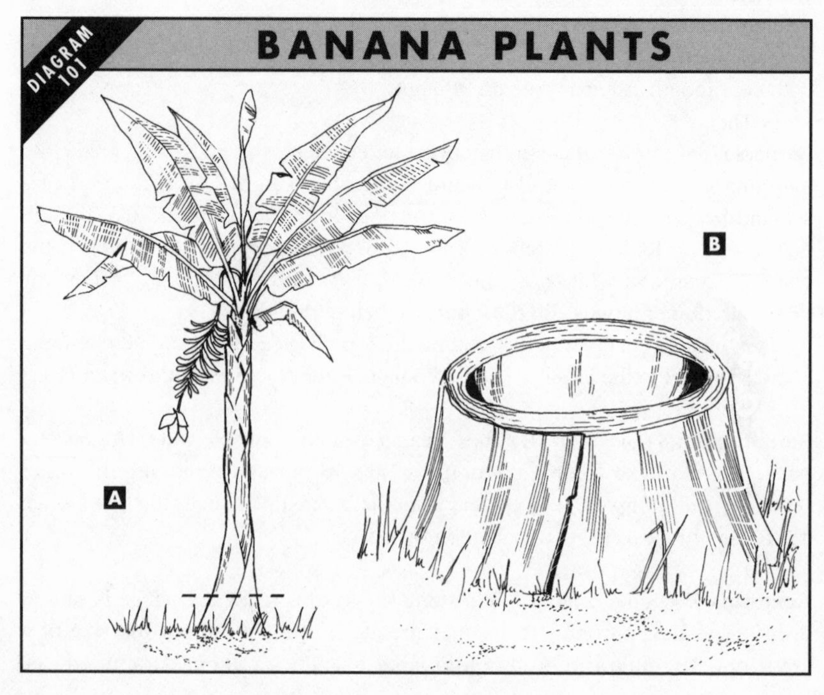
DIAGRAM 101
BANANA PLANTS
A
B

although mature coconuts contain amounts of oil, which can cause diarrhoea if taken in excess.

FOOD

The jungle is teeming with edible plants and animals. Therefore as a survivor you are well placed to eat a healthy diet in the tropics. Plants in humid regions can grow and flower all the year round, and they can grow very rapidly. Before eating any plant, however, you must put it to the edibility test (see Food Chapter), unless of course you can positively identify it.

Animals move along game trails in the jungle, and that is where you should place your traps. The animals you can hope to trap are hedgehogs, porcupines, anteaters, mice, wild pigs, deer, wild cattle, squirrels, rats and monkeys.

Although they are all edible you are strongly advised to avoid the following animals: tigers, rhinoceros, water buffalo, elephants, crocodiles, caimans and cobras (some of which spit poison into the eyes). They are all dangerous and could kill you.

Frogs Avoid all brilliantly coloured frogs: they are poisonous. Stay clear of frogs and toads that secrete a pungent odour through the skin: they are poisonous.

Reptiles They are abundant in the jungle and should be considered a food source. Treat all snakes as poisonous and kill them with a heavy blow to the back of the head.

Cats There are many members of the cat family found in tropical regions. For example, the ocelot, which has a dark-spotted, yellow-grey coat, is common in the jungles of Central and Latin America. It is small, lean and savage, weighs around 18kg (40lb) when fully grown and is approximately 91cm (36in) long.

US AIR FORCE TIPS

FISHING IN THE TROPICS

American airmen's survival training includes a course in sea fishing. The seas can provide the survivor with an abundance of fresh, nutritious food. Look for the following:

- Snails and limpets cling to rocks and seaweed above the low-water mark. Prise them off with a knife.
- Mussels form dense colonies in rock pools, on logs or at the base of boulders. NB mussels are poisonous during the summer.
- The safest fish to eat are those from the open sea or deep water beyond the reef.

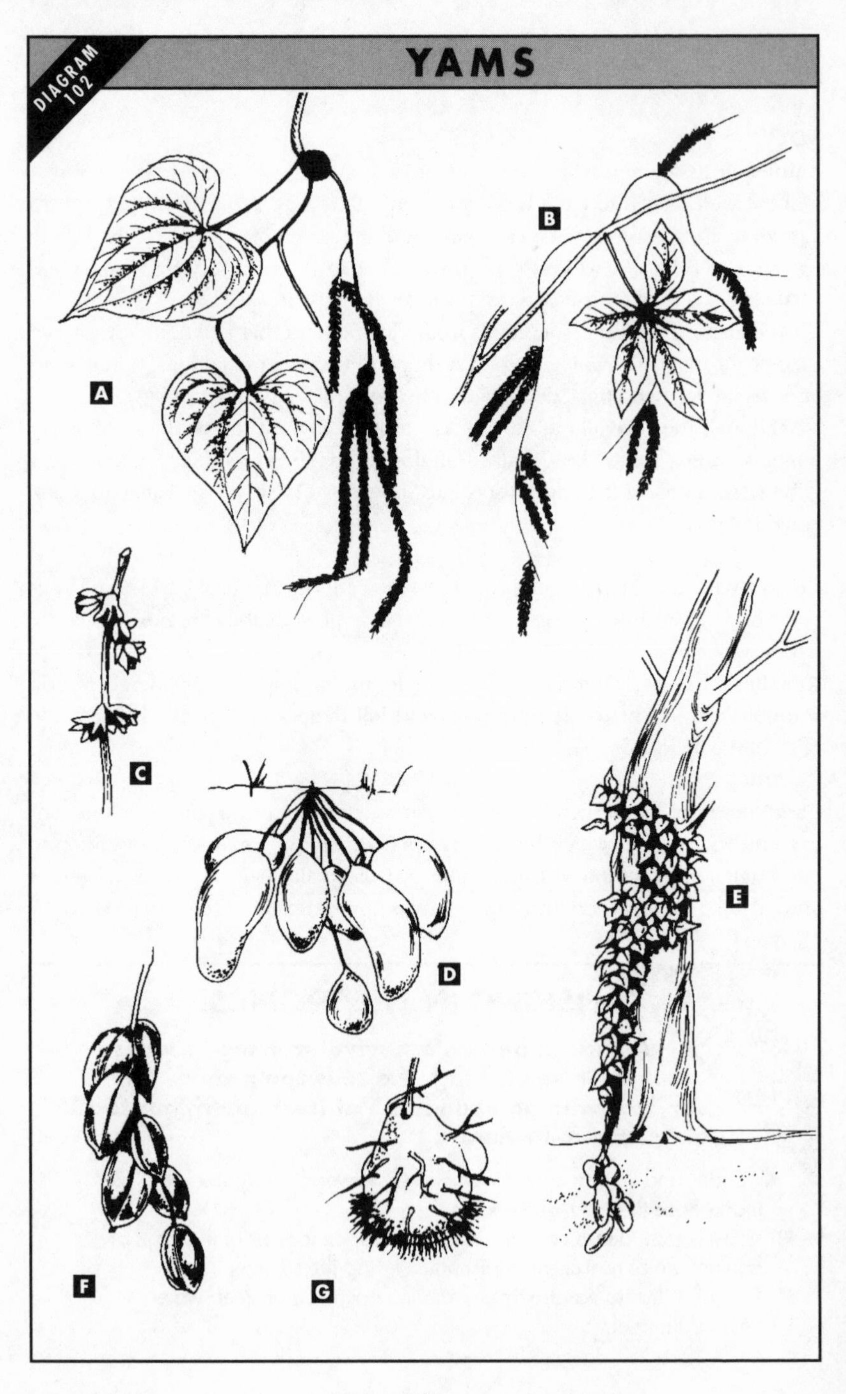
DIAGRAM 102
YAMS
A
B
C
D
E
F
G

The Leopard are also found in the tropics, though this powerful beast is not the easiest to trap. Unlike lions and tigers, it can climb trees easily. You therefore cannot take refuge up a tree if you become the hunted!

Seafood If you are near a seashore, fish crabs, lobsters, crayfish and octopuses can be a part of your diet. Try to spear or catch them before they move off into deep water.

Citrus fruit trees The leaves are usually leathery, shiny and evergreen and the flowers are usually small and white-to-purple in colour. The fruit has a leathery rind with numerous glands and is round and fleshy with several sections or slices and many seeds. Tropical fruits include *GUAVA, PERSIMMON, RAMBUTAN, DURIAN* and passion fruit.

Yams (Diagram 102) There are many varieties of yam and they occur in light forest and clearings in both tropical and subtropical regions, but the most common has a vine with a square-shaped cross section and two rows of heart-shaped leaves growing on opposite sides of the vine (A). Other varieties have five leaves on a stem, inconspicuous flowers (B) and angled seed pods (F). Yams usually have greenish flowers (C), and you can follow the vine (E) to the ground to locate the tubers (D & G). The tubers must be cooked to destroy the poisonous properties of the plant.

Coconut palms The tree is characterised by ring-like leaf scars. The leaves are leathery and can be used for shelters. The fruit grow in clusters at the top of the tree, and fall off when ripe. The coconut itself has an edible 'heart' and the flower of the coconut is also edible.

Papaya (Diagram 103) Papaya can be found in secondary jungle. The greenish or yellow fruit (A) is usually cooked. Be careful not to get the milky sap in your eyes. If you do, wash it out immediately. The fruit grows on the stem clustered under the leaves (B). It can be eaten raw or cooked after peeling. The tree (C), which can grow to a height of 1.8-6m (6-20ft) has large, dark-green, many fingered, rough-edged leaves clustered at the top (D).

Poisonous plants (Diagram 104) The following plants are poisonous and should be avoided. Some of these plants can be fatal if eaten.
NETTLE TREES (A) Appearance: resemble ordinary nettles found in temperate zones. The seeds are very poisonous.
Distribution: found throughout the tropics.
STRYCHNINE (B) Appearance: a small tree with oval leaves that grow in opposite pairs. Has white to yellowish-red fruits.

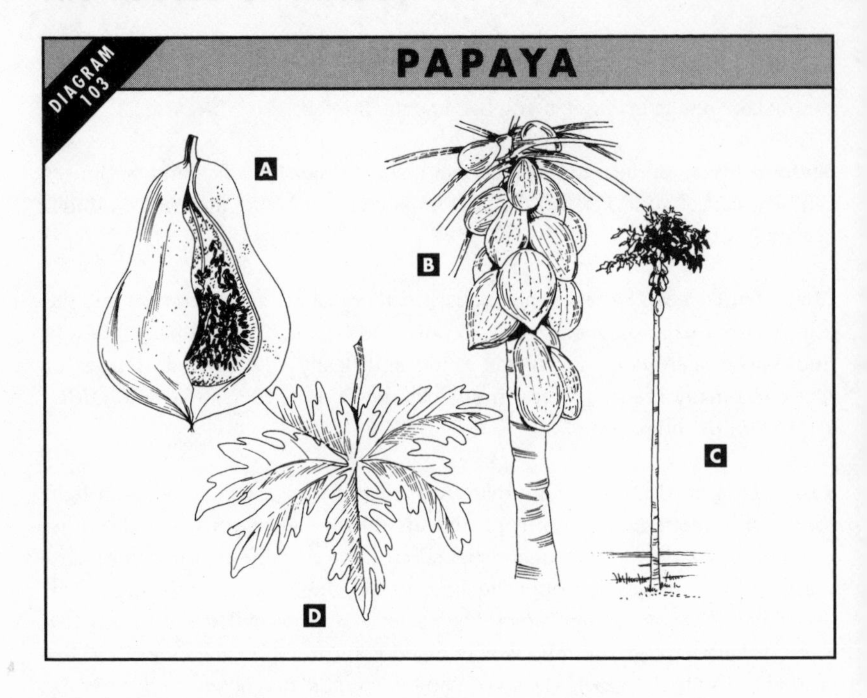

Distribution: found mainly in India, though other strychnine species are found throughout the tropics.

PHYSIC NUT (D) Appearance: large, lobed, ivy-like leaves and small greenish-yellow flowers and yellow fruits.

Distribution: found in wooded terrain throughout the tropics.

COWHAGE (E) Appearance: has spikes of hairy dull purplish flowers and brown hairy seed pods.

Distribution: scrub and light woodland throughout the tropics.

DUCHESNIA (F) Appearance: resembles an edible strawberry – red fruits and yellow flowers.

Distribution: waste ground in the warmer parts of the tropics.

PANGI (G) Appearance: a tree that has spikes of green flowers and clusters of large brownish pear-shaped fruits.

Distribution: found in the jungles of Southeast Asia, mainly Malaysia.

CASTOR BEAN (H) Appearance: spikes of yellow flowers and prickly three-sided pods.

Distribution: scrubby and waste places throughout the tropics.

WHITE MANGROVE (C) Appearance: has pale bark, pencil-like roots, yellow flowers and small white, round berries.

Distribution: grows in mangrove swamps and river estuaries in Africa, Indonesia and in tropical Australia.

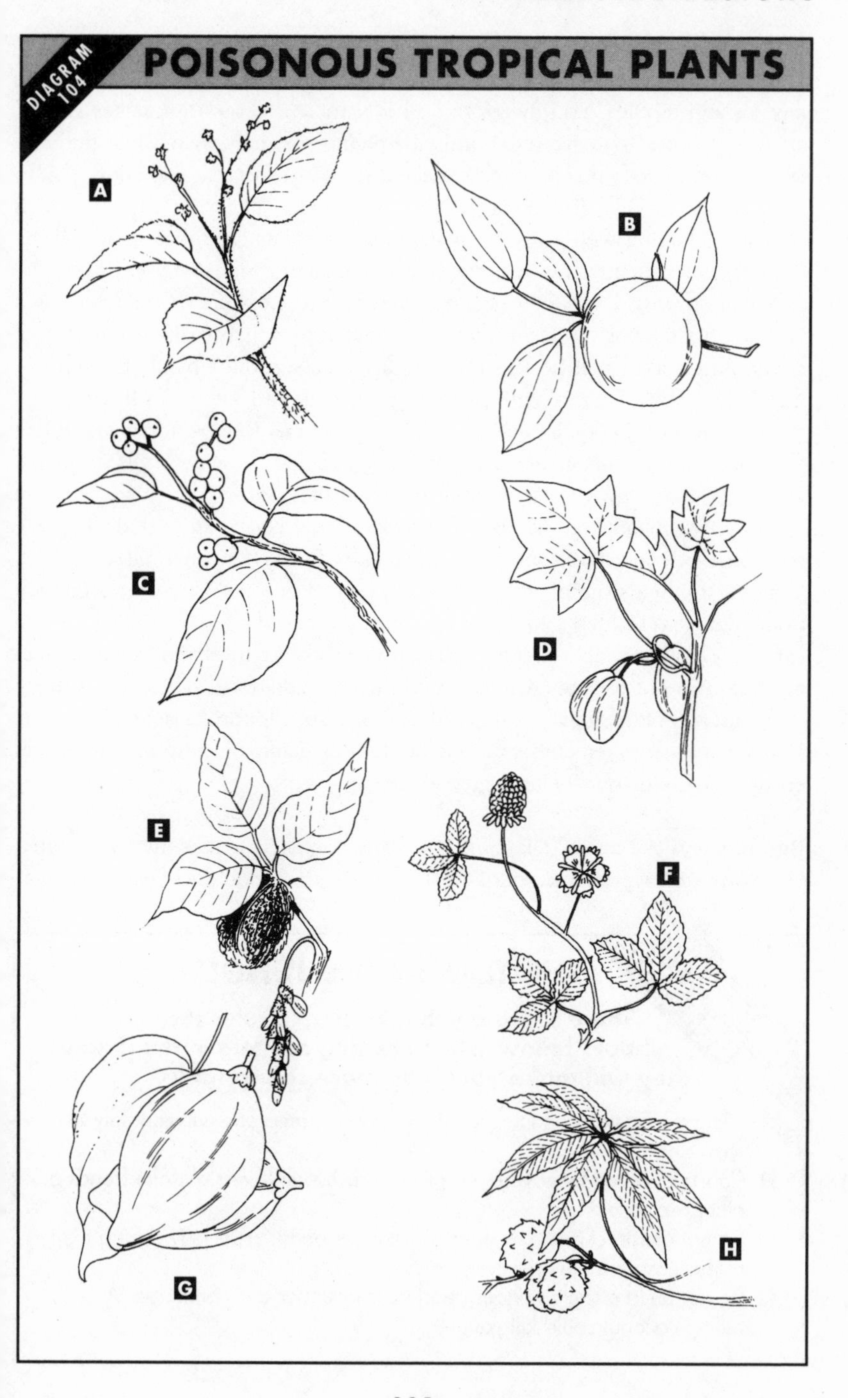
DIAGRAM 104
POISONOUS TROPICAL PLANTS
A
B
C
D
E
F
G
H

BUILDING SHELTERS

In tropical jungle and rain forests, the ground is damp and teeming with insects, leeches and reptiles. You therefore do not want to sleep on it (snakes will be attracted to your body warmth during the night – you may wake up to find one curled around your private parts!). Make a raised shelter that will allow you to sleep off the ground. If you can, you should build a shelter on a knoll or high spot in a clearing well away from standing water, where the ground will be drier, there will be fewer insects and it will be easier for signalling purposes.

When clearing a site for a shelter, remember to clear underbrush and dead vegetation. Crawling insects will not be able to approach as easily due to lack of cover, and snakes will be less likely to approach. A thick bamboo clump or matted canopy of vines will discourage insects and keep the heavy early morning dew off bedding. One more thing: remember to look above you when you have chosen your shelter site. You do not want to be below dead wood that comes crashing down in the next heavy wind or a hornets' or wasps' nest. In addition, watch out for any poisonous plants (see above and Food Chapter): some may be contact poisons which will make your stay uncomfortable.

If you are in a swamp, you will want to build a raised shelter to prevent you from getting wet. Keep a look out for four trees clustered in a rectangle which can support your weight. Cut two poles from any other trees and fasten them to the trees (do not use rotten sticks), and then lay additional poles across them. Cover the top of the frame with broad leaves or grass to form a sleeping surface. When in a swamp, remember to look out for tide marks on surrounding trees to ensure you build your shelter high enough.

Banana leaf A-frame (Diagram 105) This makes an excellent rain shelter. Construct an A-type framework and shingle it with a good thickness of palm or

CANADIAN AIR FORCE TIPS

SHELTERS IN THE JUNGLE

There are a number of simple rules that you should follow when erecting shelters in the jungle: they will make your stay more comfortable.

- *NEVER* sleep on the ground: it may be damp and will certainly be crawling with insects.
- Construct a bed by covering a pile of brush with layers of palm fronds or other broad leaves.
- Do not construct a shelter near a stream or pond, especially during the rainy season: it may get swept away.
- Do not build a shelter under dead trees or under a coconut tree. A falling coconut could kill you.

other broad-leafed plants, ie the leaves are overlapping.

Raised platform shelters (Diagrams 106 and 107) These shelters have many variations. The poles should be lashed together and crosspieces secured to form the platform on which material mattresses can be made. Try to make the roof waterproof with thatching laid from bottom to top in a thick shingle fashion (it also helps to have a mosquito net).

You can use split bamboo to make roofing: cut the stem in half and lay them alternately to interlock with each other. You can also flatten split bamboo

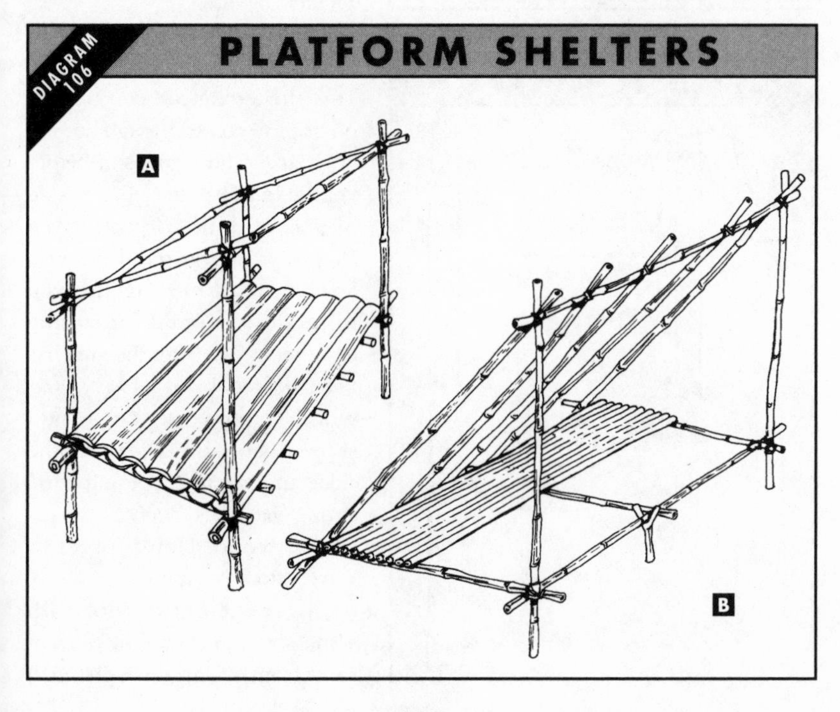

and use it for lining walls or shelving.

Diagram 106A shows the framework for a raised platform shelter with a banana log bed; 106B illustrates the framework for a raised lean-to shelter.

Diagram 107 shows a paraplatform shelter: a shelter with a wooden framework and roof made from parachute material or a poncho (don't forget to tie the roof securely).

Hammock (Diagram 108) A hammock can be made quickly if you have a poncho or similar type of material and rope. A hammock may be tied between two trees or three or more for greater stability.

Seashore shelter (Diagram 109) This shelter can be constructed on tropical coasts, though be sure how far the tide comes in before you build it. Dig into the lee side of a sand dune to protect the shelter from the wind. Clear a level area large enough for you to lie down in and for storing equipment (A). After the area has been cleared, build a heavy driftwood framework which will support the sand. Then wall the sides and top (B). You must use strong materials, such as boards or driftwood. Don't forget to leave a door opening.

Cover the entire roof with some sort of material to prevent sand from sifting through small holes in the walls and roof (C). This material should be fairly thick and hard-wearing. Cover the roof with 15-30cm (6-12in) of sand to provide protection from the wind and moisture. Lastly, make a door for the shelter (D). Check the shelter regularly for signs of wear and tear.

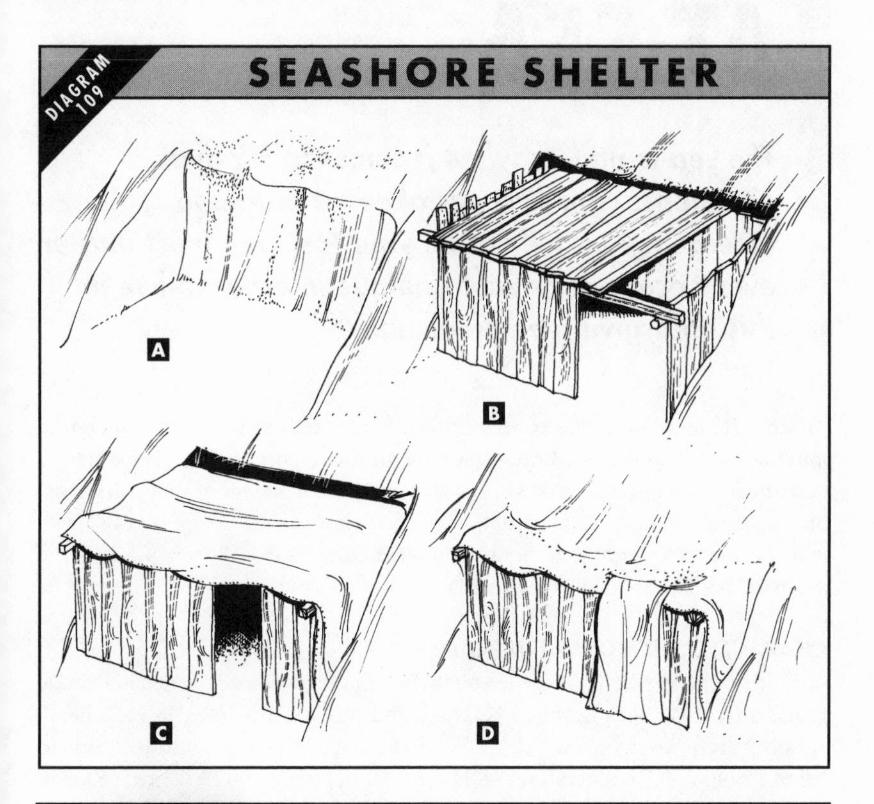

BRITISH SAS TIPS

MATERIALS FOR JUNGLE SHELTERS

There is an abundance of shelter-building material in the jungle. Know what it is and how to use it, but also be aware of its dangers.

- Atap, which has barbs at each leaf tip, is a vine that can be used to make shelters. Split each leaf from the tip and layer it on frames.
- Three-lobed leaves can be thatched on a frame.
- Elephant grass is large and can be woven on a frame.
- Bamboo can be used for pole supports, flooring, roofing and walls.

Take care when collecting bamboo: it grows in clusters and some stems are under tension. They can fly off and cause serious injury.

SURVIVAL AT SEA

The sea is pitiless when it comes to survival situations: your first mistake in a maritime emergency is likely to be your last. You must master every ocean survival technique in order to live in this unforgiving environment.

Around 71 per cent of the earth's surface is covered by water. It is therefore imperative that you learn how to survive in this environment. Man's natural domain is land, and that makes survival at sea for any length of time a formidable challenge. In particular, acquiring drinking water and food are serious problems for the survivor at sea, though the other dangers that the sea poses to the survivor should not be underestimated.

CHARACTERISTICS OF THE TERRAIN

Around the poles, the limit of the solidly frozen arctic ice pack varies in latitude from 65 to 75 degrees between February and August. In winter, in each hemisphere, there are cyclonic storms characterised by snow, winds of up to 64km/hr (40mph), temperatures as low as -50 degrees C (-122 degrees F), and gale-force winds. In summer, there are periods of calm or days with light winds. Skies are overcast, there are dense fog banks during calm periods, and rain or drizzle may last for weeks.

Around the 40 degrees latitudes in each hemisphere there is generally fair, clear weather, with temperatures of around 10 degrees C (50 degrees F) in the winter and around 21 degrees C (70 degrees F) in summer. Below a 25-degree latitude, in the heart of the trade wind belt, winds of 8-24km/hr (5-15mph) are normal. There is little difference between summer and winter, with temperatures being in the 21-27 degrees C (70-80 degrees F) range.

Between five degrees north and five degrees south in the Atlantic, Pacific and Indian Oceans, there is an equatorial trough of low pressure where there are no prevailing surface winds. As a result, there are shifting winds and calms. Solar heat results in violent thunderstorms.

Waterspouts (the marine equivalent of tornados) are common off the Atlantic and Gulf coasts of the USA and along the coasts of China and Japan. Hurricanes and typhoons occur in the warm areas of all oceans during the summer and autumn. They can last for up to two weeks.

Salinity The average salinity of the seas is around 3.5 per cent, though higher values occur at or near the surface in areas where high temperatures and strong, dry winds favour evaporation.. The highest salinities occur in semi-landlocked seas at mid-latitudes, such as the Red Sea, the Persian Gulf and the Mediterranean Sea. The salt in the water makes it undrinkable and can aggravate open sores and cuts (see below).

IMMEDIATE ACTIONS

When you hit the water, try to make your way to a raft. If none is available, try to find a large piece of floating debris to cling to. Try to stay calm: a relaxed body will stay afloat. Floating on your back requires less energy then swimming face-down. Alternatively, float face-down on the surface with your arms outstretched and legs pointing towards the bottom. To breathe, raise your head and place your arms in the outstretched position again.

Cold water If the water is cold, you risk dying of hypothermia if fully immersed. You must get into a raft and insulate your body from the cold bottom of the raft. If there isn't a raft, keep still and assume the Heat Escape Lessening Posture (HELP) position (Diagram 110): this will increase your survival time. Around 50 per cent of the body's heat is lost from the head, therefore try to keep your head out of the water.

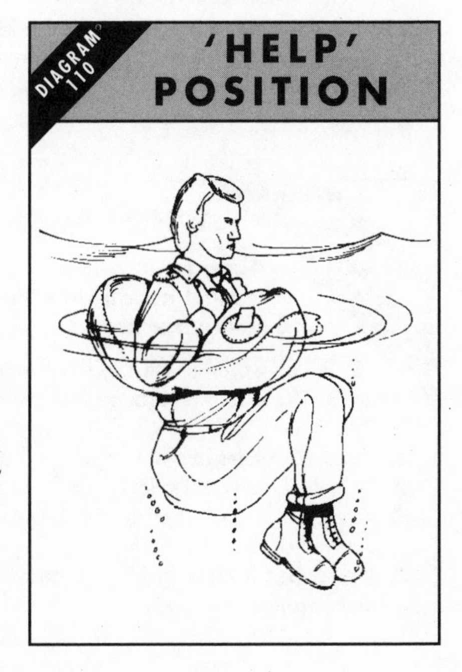

If there are several of you in the water, huddle close in a circle to preserve body heat (Diagram 111.). However, note that these measures are only temporary – *YOU MUST GET OUT OF THE WATER.*

Clothing at sea If you are adrift in a cold ocean, you must try to stay dry and keep warm (use a wind

screen to decrease the cooling effects of the wind). If possible, remove, wring out and replace outer garments or change into dry clothing. If any survivors have dry clothes, they should share them with those who are wet. Those who are wet should be given the most sheltered positions in the raft, and let them warm their hands and feet against those who are dry. If possible, give extra water rations to those suffering from cold exposure.

Survivors use any extra clothing they have (drape it around the shoulders), and cover the floor of the raft with any spare material to help keep it insulated. Huddle together on the floor of the raft and spread extra sail or parachute material over the group.

You should exercise fingers, toes, legs, arms, shoulders and buttocks to prevent muscle spasms, help keep the body warm and prevent medical problems, such as constipation. Put your hands under your armpits then raise your feet slightly off the ground and keep them up for a minute or two. Try to exercise at least twice a day.

BRITISH SAS TIPS

WHEN ABANDONING SHIP

Abandoning ship is a frightening experience, but you must act quickly. Follow these SAS guidelines and save your life.

- Put on warm, preferably woollen, clothing, including hat and gloves. Wrap a towel around your neck.
- Take a torch.
- Grab chocolates and boiled sweets if possible.
- Do not inflate your life jacket until you leave the ship.
- When jumping overboard, first throw something (anything wooden) that floats and jump close to it.
- Air trapped in clothing will help buoyancy: do not take off your clothes in the water.

US ARMY TIPS

RECOMMENDED SWIMMING STROKES FOR SURVIVORS

You must try to conserve your strength when you are in the water. Use a variety of these swimming strokes when you have survived a disaster:

- Dog paddle: good for when you are clothed or wearing a life jacket.
- Breast stroke: good for swimming underwater or in rough seas.
- Side stroke: a useful relief stroke because only one arm is needed to maintain momentum and buoyancy.
- Back stroke: another good relief stroke. It relieves the muscles that are used for other strokes.

Picking up survivors If you are in a life raft and are rescuing people in the water, try to throw them a line with a life belt attached. Alternatively, send a swimmer out with a line attached to a flotation device. If you are in the water rescuing people, approach them from behind to avoid getting kicked, grabbed or scratched. Grab them by the back of their life jacket or hold them under the chin and then use a sidestroke to drag them to the raft. Try to reassure them as you do so. *DO NOT UNDERESTIMATE THE STRENGTH OF A PERSON IN A STATE OF PANIC IN THE WATER.*

CANADIAN AIR FORCE TIPS

LIFE RAFT ROUTINE

When adrift in a life raft you must undertake the following practices to keep the raft seaworthy and maintain morale.

- Repair the raft as soon as it gets damaged with patches or other materials (check the raft frequently for any damage).
- Bleed off air if the raft expands in hot weather.
- Top up air if the raft contracts in cold weather or at night.
- Relax and try to keep your mind occupied: keep a log.
- If in a group, form a team and ensure everyone has a job to do. This will help to pass the time and lessen the possibility of seasickness.

MOVEMENT

You must remember one thing when afloat in the ocean: your raft will be at the mercy of winds and currents. Currents flow in a clockwise direction in the Northern Hemisphere and anti-clockwise in the Southern Hemisphere. Sea currents travel at speeds of less than 8km/hr (5mph), so movement is going to

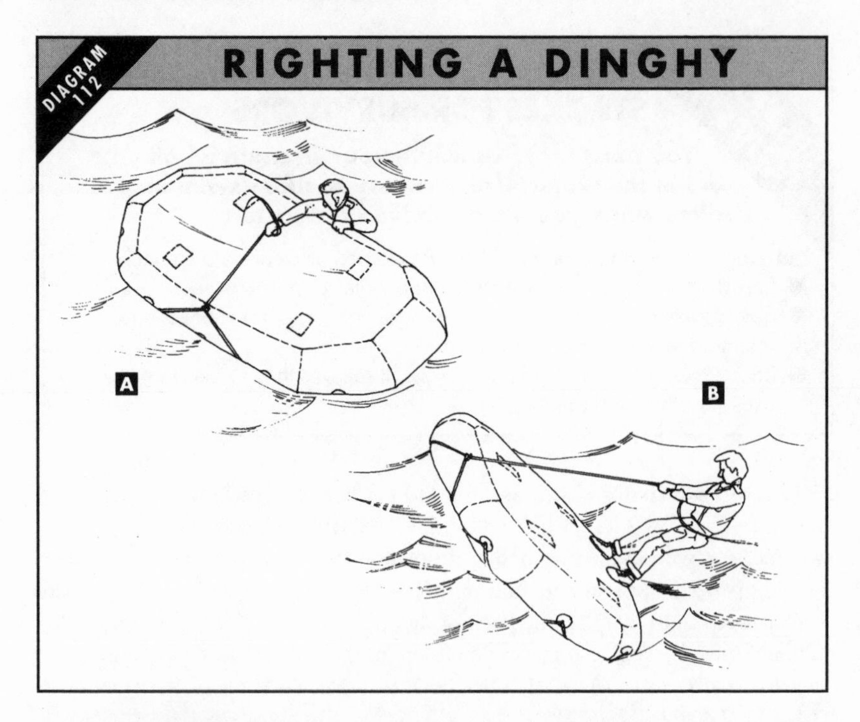

RIGHTING AN INFLATABLE DINGHY (DIAGRAM 112)

BRITISH SAS TIPS

A capsized dinghy need not be a catastrophe in a survival situation at sea: you can right it easily if you know how.

- Grab the righting line from the opposite side (A).
- Brace your feet against the dinghy and pull.
- The dinghy should rise up and over, and will pull you temporarily out of the water (B).
- This procedure requires more effort in heavy seas or high winds.

be very slow. In areas where warm and cold currents meet there will often be storms, dense fog, high winds and heavy seas. These will make movement difficult and dangerous.

Winds and waves can aid raft travel. Winds blow in an easterly direction in tropical areas, the so-called trade winds, and from the west in higher latitudes. To take advantage of the wind you will need a sail. If the raft doesn't have one, improvise one from a poncho or other piece of material.

BRITISH SAS TIPS

INDICATORS THAT LAND IS NEARBY

As a survivor at sea, your number one priority is to reach land. Keep a lookout for the following objects and signs that can point to land.

- Clouds. Cumulus clouds (fluffy, white ones) in a clear sky are likely to have been formed over land. In tropical areas, a green tint on the underside of clouds is caused by the reflection of sunlight from the shallow water over coral reefs.
- Birds usually fly from land before midday and return to it in the late afternoon. Beware of lone birds though: they may just be disorientated.
- Coconuts, driftwood and drifting vegetation can sometimes indicate that land is near.
- A change in the direction of sea movement may be caused by the tide around an island.
- Water that is muddy with silt has probably come from the mouth of a large river that is nearby.
- Deep water is dark-green or dark-blue: a lighter colour indicates shallow water and perhaps land.

Waves can be both an asset and a hazard. The size of waves is usually dependent upon the severity of the wind. Waves will only move a raft a few centimetres at a time under normal conditions, so they are of little use as a means of propulsion. However, they are an asset when you are searching for shallow areas or land. Ocean waves break when they enter shallow water or when they encounter an obstruction. Use breaking waves as an aid to making a landfall. Waves can also capsize a raft in bad weather, or fill it with water.

To move or stay put? If an SOS has been sent (see Signalling and Navigation Chapter), or you know you are in regular shipping lanes, you should stay in the same location for up to 72 hours. However, if you are off the shipping lanes and didn't manage to send a signal, then you should get underway as soon as possible to take advantage of your fitness and stamina. Head in the direction of land. If there is no land nearby, try to work out where the nearest shipping lane is and head in that direction.

Signalling Use flares and dye markers (which spread brightly coloured dye in the sea) if you have them to attract the attention of a ship or aircraft. If you do not have any signalling equipment, attract attention to yourself by waving clothing and other materials, brightly coloured if possible. Sea markers should only be used in daylight (they normally last for around three hours). A mirror or reflective surface can be used for long-range signalling.

All flares should be handled carefully. Keep them dry and secure, and when firing point them upwards and away from yourself and anyone else in the raft. Use them only when you are sure that they will be seen. A shiny, reflective surface is also an excellent way of attracting attention to yourself.

If you have a radio transmitter in your life raft, it will have preset frequencies at 121.5 and 243 megacycles, and will have a range of around 32km (20 miles). Try to transmit at frequent intervals but be careful not to run down the batteries (if you have a watch you can use it to time signals at regular intervals – remember to keep all watches dry).

Getting ashore Once you have sighted land you will obviously make for it without delay.

If you are swimming, wear your shoes and at least one thickness of clothing. Use side or breast stroke to conserve your strength. Water is calmer in the lee of a heavy growth of seaweed. *DO NOT* swim through it: crawl over the top by grasping the vegetation

If you are in a raft, choose your landing point carefully. Do not land when the sun is low and straight in front of you, and avoid coral reefs and rocky cliffs. Similarly, stay well clear of rip currents (a strong surface current) or strong tidal

SEAL TIPS

SWIMMING ASHORE

In the water and approaching a shore line? Though you are in sight of land you are also in potential danger. Employ SEAL tactics to land safely.

- Ride in on the back of a small wave by swimming forward with it.
- In high waves, swim towards the shore in the trough between waves. Face and submerge beneath waves, then swim forward in the next trough.
- If caught in the undertow of a large wave, let it pass over and then push off the bottom with your feet or swim to the surface if in deep water.
- When landing on a rocky shore, aim for the place where the waves rush up onto the rocks, not where they explode with a high white spray.
- To land, advance behind a large wave into the breakers. Face the shore with your feet in front, 0.5-1m (2-3ft) lower than your head. In this way your feet will absorb shocks when you land or hit submerged rocks or reefs, and you will not get injured.
- If you don't reach shore behind the wave you have selected, swim using your hands only. Adopt a sitting position as the next wave approaches and carries you to shore.

currents. Use oars and paddles if you have them and adjust the sea anchor to keep a strain on the anchor line. This will prevent the sea from throwing the stern around and will keep the raft pointed towards the shore.

In heavy surf try to avoid meeting a large wave at the precise moment it breaks. As you near the beach, try to tide the raft on the crest of a wave. Do not jump out until the raft has grounded, and then get out quickly and pull it ashore. If you come across sea ice, try to land on large, stable floes. Use the oars to keep the raft's edge away from ice edges, and remember that when you are on a floe it may break up at any time. Keep the raft ready.

DANGERS

The many dangers at sea, such as hostile fish, the threat of starvation and thirst and the cold and wet, can all be dealt with to varying degrees, though remember that you rarely get a second chance at sea. Stay alert!

Sharks are scavengers and live in almost all seas and oceans. They feed more actively at night, and especially at dawn and dusk, and after dark they move towards the surface and into shallow waters. They are attracted to rubbish, body wastes and blood, and also by weak fluttery movements similar to a wounded fish. A shark cannot stop suddenly or turn quickly in a tight circle, and it will rarely jump out of the water to take food. For this reason, people on rafts are relatively safe from attack, unless they dangle their legs or arms in the water.

If you are on a raft or dinghy, do not fish when sharks are around and do not throw any waste overboard. If it looks as though a shark is going to attack, try to jab its snout with a pole or pebble.

If you catch a small shark while fishing, try to haul it in to the side, pull its head clear and then club it hard before hauling it aboard. You must ensure it is stunned before approaching it and finishing it off. If you catch a large shark, cut the line and let it go: it could damage your vessel and injure you. In addition, its threshing will attract other sharks.

The main types of shark that have been known to attack humans are listed below, but be aware that all sharks, because of their sharp teeth and aggressive feeding habits, must be considered potentially dangerous. One more thing: there is no relationship between the size of a shark and the risk of attack.

NURSE Appearance: greyish coloured on top, white underneath, very heavily built and large-finned.
Size: around 4m (13ft).
Temperament and habits: aggressive, often found close inshore.
Distribution: around eastern Australia.
BULL SHARK Appearance: grey on top, white underneath with a stout body.
Size: up to 4m (12ft).
Temperament and habits: likes shallow waters and rivers. Aggressive.

Distribution: tropical west Atlantic with close relatives off southern Africa and in the Indian Ocean.
HAMMERHEAD SHARK Appearance: flat, hammer-like head, long body.
Size: up to 6m (18ft).
Temperament: can be aggressive.
Distribution: tropical and subtropical waters.
TIGER SHARK Appearance: grey on top, white underneath with a very wide head and jaws.
Size: 3-3.5m (12-13.5ft).
Temperament and habits: often found close inshore, can be dangerous.
Distribution: tropical and subtropical waters.
MAKO SHARK Appearance: ultramarine blue on top and creamy-white underneath. Brightly coloured.
Size: 2-3m (6-9ft).
Temperament and habits: can swim very quickly, sometimes leaps from the water when agitated.
Distribution: warm temperate waters.
GREAT WHITE SHARK Appearance: grey on top, white underneath, thick body, conical snout and black eyes.
Size: up to 6m (18ft).
Temperament: very aggressive.

SEAL TIPS

ANTI-SHARK MEASURES IF YOU ARE IN THE WATER

SEAL team members are often on operations in shark-infested waters. They have tried and tested anti-shark measures.

- If you are in a group in the water, bunch together and form a tight circle for added protection.
- Face outwards and keep a lookout for sharks.
- Ward off an attack by kicking or stiff-arming a shark. Only use your hands as a last resort. Use a hard object, such as a knife, and aim for the snout, gills or eyes.
- Urinate in short, sharp bursts and allow it to dissipate between spurts. Collect faeces and throw as far away from you as possible. Try to re-swallow vomit or, failing that, throw it as far away as possible.
- Stay as quiet as possible and float to save energy.
- If you have to swim, use strong regular strokes, not frantic irregular movements (these will make sharks believe you are a wounded fish).
- Do not swim directly away from a shark; rather, face it and swim to one side, using strong, rhythmic movements.

Distribution: found in all the warm and temperate oceans of the world, but especially off southern Africa, east and west of North America and southern Australia and New Zealand.
COW SHARK Appearance: sandy-grey with dark spots.
Size: up to 3m (9ft).
Temperament and habits: swims close to the surface, aggressive.
Distribution: tropical and subtropical waters.
SAND SHARK Appearance: white underneath, mouse-grey on top with yellow spots, hence its name.
Size: up to 3m (9ft).
Temperament and habits: aggressive when provoked.
Distribution: tropical and subtropical waters.
SNAGGLETOOTH Appearance: coloured golden brown or grey.
Size: 2.5m (6ft).
Temperament and habits: can be found in shallow waters, can be aggressive.
Distribution: tropical waters.
SILVERTIP SHARK Appearance: charcoal, white tips on fins.
Size: 3m (9ft).
Temperament and habits: fast and bold, potentially dangerous. Plentiful around reefs and islands.
Distribution: tropical and subtropical waters
GREY REEF SHARK Appearance: grey, tail edged with black.
Size: 2.5m (6ft).
Temperament and habits: curious but not aggressive.
Distribution: tropical waters.
COPPER SHARK Appearance: golden brown and cream underneath.
Size: 3m (10ft).
Temperament and habits: can be very aggressive.
Distribution: tropical and subtropical waters.
BULL SHARK Appearance: grey on top, off-white underneath.
Size: 3.5m (11.5ft).
Temperament and habits: dangerous: the most feared of tropical sharks.
Distribution: the tropics: will swim up rivers.
BLUE SHARK Appearance: brilliant blue on top, white underneath.
Size: 4m (13ft).
Temperament and habits: one of the most dangerous sharks in the ocean; responsible for many human fatalities and .
Distribution: worldwide in tropical and temperate waters.
Remember that not all fish you see with fins are sharks. For example, dolphins and porpoises have long snouts and are enemies of sharks, while being no threat to humans. Giant rays are present in tropical waters. Sometimes they curl up the tips of their fins, and this may look like two sharks swimming side by

BRITISH SAS TIPS

AGGRESSIVE FISH

Take note of the following fish: they may be dangerous in a survival situation. Above all, learn to recognise them and give them a wide berth.

- The Sea Bass is curious and bold. It is found around rocks, caverns, old wrecks and caves. Avoid these areas.
- Moray eels live in holes and crevices. They are aggressive if disturbed. If you are attacked, try to cut off its head since they retain their grip until dead (this may be difficult because they are slippery to the touch).
- Sea snakes are deadly poisonous. They are unlikely to bite but stay well clear of them.

side. All rays are harmless to survivors in deep water, but in shallow waters some are dangerous if stepped on: they have venomous spikes in the tail.

Killer whales usually hunt in packs of up to 40 and sometimes attack anything that floats or swims. Get out of the water if killer whales are nearby if you can. On ice, do not stand near seals, as the whale may mistake you for one and try to eat you. However, it must be stressed that the chances of being eaten by a killer whale are very low.

Barracuda Found in tropical and subtropical seas, this fish is attracted by anything that enters the water, and especially bright objects. Barracuda are very

US AIR FORCE TIPS

DEALING WITH JELLYFISH STINGS

A jellyfish sting can be extremely painful, and in extreme cases can be fatal. Follow these procedures for dealing with them.

- Remove tentacles or other matter from the skin immediately.
- Use clothing, seaweed or other material to do this.
- *DO NOT* rub the wound with anything, especially sand, as this may result in activating the stinging cells.
- *DO NOT* suck the wound.
- Try to alleviate the poison effects: try soap, lemon juice, plant juices, baking powder or urine, which has an ammonia content.
- Artificial respiration and CPR may be required (see First Aid Chapter).
- Since stings may result in convulsions, it is important to get the victim out of the water.

fast and often roam in shoals. They are dangerous when there is blood in the water. Try to wear dark-coloured clothing if you have to enter the water, and never wear anything shiny.

Jellyfish There are many different kinds of jellyfish. The largest can be of 1.8m (6ft) in diameter, with tentacles hanging down to a depth of up to 30m (100ft). These tentacles contain stinging cells that can inflict serious injury on the survivor. One of the deadliest is the sea wasp, which can cause death in as little as 30 seconds, though around three hours is normal.

All jellyfish should be given a wide berth, especially since their tentacles may trail a long way from the body. Following a storm in tropical areas where large numbers of jellyfish are present, you may be stung by pieces of floating tentacles that have been removed from the fish during the storm. Jellyfish washed up on a shore may look dead, but many can still inflict painful injuries. In general, try to get out of the water when jellyfish are present.

Poisonous fish There are many reef fish that are dangerous to eat, especially bottom dwellers and feeders. For example, many fish from reefs and lagoons are poisonous to eat. If in doubt do not eat. Fish toxins are water soluble and tasteless, therefore you cannot use the taste test (see Food Chapter) to prove edibility, and no amount of cooking will neutralise them. In addition, birds can be immune to these poisons: so just because you see a bird eating a fish, do not think that it is safe for you to eat it.

Ingestion of fish toxins can result in death in the worst cases. As soon as any symptoms arise – numbness, itching, reversal of temperature sensations, nausea – induce vomiting by administering saltwater or the whites of eggs (give victim a laxative if available). You may have to perform a cricothyrotomy (see First Aid Chapter) if the victim foams at the mouth and shows signs of respiratory distress. Cool showers may relieve itching. Be prepared to treat any other symptoms as and when they arise.

There are other fish that are dangerous to touch, either because they have venomous spines, e.g.stonefish, or because they have poisonous barbs in their tails, e.g. rays. The spines inject a venom that is excruciatingly painful. Keep shoes on at all times when you are walking through saltwater, and use a stick, not your hands, to explore sand, rocks and holes.

The survivor should be aware of the following poisonous fish:

RABBITFISH Appearance: bright yellow with tiny blue spots.
Size: around 25-30cm (10-12in).
Threats: has venomous spines along its fins.
Distribution: reefs in the Pacific and Indian Oceans.
SURGEONFISH Appearance: very brightly coloured (usually black, blue and yellow), small-mouthed.

Size: 20-25cm (8-10in).
Threats: spines on the sides of the tail can inflict severe wounds.
Distribution: all tropical waters.
SCORPIONFISH Appearance: reddish, striped, with long, wavy spin spines.
Size: 3075cm (12-30in).
Threats: its sting is intensely painful.
Distribution: reefs in the tropical Indian and Pacific Oceans.
VENOMOUS TOADFISH Appearance: dull-green coloured and large-mouthed.
Size: 30-40cm (7-10in).
Threats: lie buried in the sand and have sharp, poisonous spines on their back.
Distribution: tropical waters off the coasts of Central and South America.
STONEFISH Appearance: drab coloured and lumpy shaped.
Size: around 40cm (16in).
Threats: dorsal spines inject a venom that is very painful, sometimes fatal.
Distribution: tropical Pacific and Indian Oceans.

SEAL TIPS

TREATMENT OF POISONS INJECTED VIA FISH SPINES

Follow this effective SEAL treatment for dealing with poisons injected via the spines of tropical and subtropical fish.

- Irrigate wound with water.
- Make a small cut across the wound and apply suction. Even if no cut is made, suction should be applied to remove as much venom as possible.
- Suck out as much poison as possible.
- Soak the injured part in hot water for 30 minutes-one hour. The water should be as hot as is bearable. If the wound is on the face or body, use hot compresses.
- Clean the wound after soaking.
- Cover the wound area with antiseptic and a clean, sterile dressing.
- You may have to treat for shock (see First Aid Chapter).

Marine snails and slugs As a survivor you may come into contact with these gastropods when you are crossing coral reefs and sandy shores. Avoid them: the animal is capable of injecting its poison via a barb into the flesh of a victim. The sting made by a cone shell is a puncture-type wound.

When you are stung, the area around the wound will turn blue, swell, become numb, sting and burn. The degree of pain varies from person to person, though in all victims the numbness and tingling sensation around the wound quickly spread through the whole body. This can be followed in a matter of hours by complete muscular paralysis, coma and death.

There is no specific treatment for this kind of injury. The best that you can do is apply hot towels or soak the wound in hot water to relieve the pain. You may have to administer artificial respiration.

Octopus These creatures like to hide in holes or underwater caves. You should avoid these areas. The sharp parrot-like beak of the octopus makes two small puncture wounds when it bites you, and into these wounds it injects a toxic venom. You will feel immediate pain in the form of burning, itching or stinging. Bleeding will be profuse and the area around the wound may swell, turn red and feel hot. Especially dangerous is the blue-ringed octopus, found off eastern Australia, whose bite is and lethal.

There in no known cure for the bite of the blue-ringed octopus. For other octopus bites, treat as for shock (see First Aid Chapter). Stop the bleeding and clean the wound area, since there may be more venomous saliva around the wound. Then treat symptoms as they arise.

Medical problems In a raft, do not expose yourself to the sun and wind needlessly. Keep a layer of clothing on at all times, especially on your head. If you are very hot, dip your clothing in the sea, wring it out and put it back on (this is an extreme measure). Place any injured persons on the floor of the raft and make them comfortable. Try to keep them as warm or cool, depending on the climate you are in, and as dry as possible.

Wear sunglasses or eye shields to protect your eyes from the sun's glare. Be especially careful about reflection off the water, which intensifies the sun's rays. Do not rub sore eyes: apply an antiseptic cream to the eye lids and bandage lightly instead.

For parched lips and cracked skin: apply sun cream or Vaseline and do not lick your lips. Cover dry skin to prevent further drying. Prolonged exposure to saltwater may produce cold sores. Try to keep your clothing as dry as possible.

US ARMY TIPS

TREATING SEASICKNESS

Seasickness is not a joke: it can seriously weaken sufferers and lower the morale of other survivors. Deal with it the US Army way.

- Wash patient and raft to remove the sight and small of vomit.
- Do not let patient have food until his or her nausea has gone.
- Make patient lie down and rest.
- If available, administer seasickness pills.

Clean the sores and apply antiseptic cream. Large sores should be covered with a dressing (change dressings regularly).

FINDING WATER AND FOOD

Water The lack of drinking water is a major problem for the survivor at sea. Whatever water you have must be rationed at once. Never relax the ration: you do not know how long you will need it. *NEVER DRINK SEA WATER OR URINE.*

Water sources If your raft has a de-salter kit or a solar still, use according to the instructions provided. You should employ every means available to catch and store rain water. Use the canopy on the raft to collect rain (wash it in sea water prior to collecting rain, the slight salt contamination will cause you no ill effects). In addition, rig the canvas at night to catch any dew (see Water Chapter).

BRITISH SAS TIPS

WATER RATIONS

At sea, water is your most precious commodity. Follow these water ration rules strictly to increase your chances of survival.

- Day 1. Give no water: the body can make use of its own water reserves. Be strict with this rule.
- Days 2-4. Give 400cc (14oz) if available.
- Day 5 onwards. Give 55-225cc (2-8oz) daily, depending on water availability and climate.

CANADIAN AIR FORCE TIPS

REDUCING FLUID LOSS

Follow strictly these Canadian Air Force guidelines for reducing your body's overall use and loss of fluids.

- If you do not have any water do not eat.
- In hot climates reduce loss of body fluids through perspiration by remaining inactive.
- Brush dried salt off the body with a dry cloth.
- Try to sleep and rest as much as possible to minimise fluid loss.
- Try not to get seasick. Vomiting means losing valuable fluids. Relax and focus your mind on other tasks.
- Do not drink alcohol: it dehydrates the body.
- Do not smoke: it increases thirst.
- Suck on a button to stimulate saliva and thus reduce the desire to drink.

When it is raining, drink as much rain water as possible while you are collecting it. However, remember to drink slowly. If you have been on a water ration, gulping down a lot of fluids will make you vomit.

Sea ice can be a source of drinking water, but only use old sea ice (it is blue-grey in colour and has rounded contours): it has very little salinity. Pools on old sea ice in summer can be drunk, though ensure the water is not from wave splashes (they wil be salty and undrinkable).

You can drink the aqueous fluid found along the spines and in the eyes of large fish. To get at the fluid along the spine cut the fish in half. Suck the eye. Do not drink any other of the fish's body fluids: they are rich in protein and will use your body's water supply to digest them.

Food The amount of food you should eat in a survival situation is in direct proportion to the amount of drinking water you have. Do not eat food if you have no water: it will only deplete your body fluids. If you have emergency rations only use them when they are really needed. Try to live off caught food if you can. Remember, you do not know how long you will be adrift.

Fish will be your main food source (flying fish may even jump into your raft!). In the open sea, out of sight of land, fish are generally safe to eat. However, do not eat fish that are brightly coloured, covered with bristles or spines, or those which puff up or have parrot-like mouths or humanoid teeth. Similarly, avoid fish eggs in clusters or clumps: these will be poisonous.

BRITISH SBS TIPS

FISHING AT SEA

The men of the Special Boat Squadron are renowned aquatic soldiers. Follow their advice when fishing at sea.

- Do not touch the fishing line with bare hands when reeling in, and never wrap it around your hands or tie it to an inflatable dinghy. The salt on it makes a sharp cutting edge that is a danger to your hands and the raft.
- If you have gloves wear them when handling fish: that way, you will not get fins or fish teeth in your skin.
- Pass a net under your raft from one end to the other – fish and turtles are attracted to the shade under your raft (this requires at least two people to perform).
- Use a torch to attract fish at night.
- Use improvised hooks – small folding knives, wire, pieces of jagged metal – and small bright objects, such as buckles, for bait.
- Use offal from caught fish for bait.

Birds All sea birds are potential food. They will be attracted to your raft as a perching place. Wait till they settle, you may be able to grab them if they are exhausted by flying in bad weather. In addition, you can trap birds by trailing lines in the water with hooks baited with fish.

Seaweed You can collect seaweed around shore lines and in mid-ocean. Seaweed is a rich source of minerals, though it can act as a violent laxative if your stomach is not accustomed to it. Remember that seaweed absorbs fluids when your body is digesting it, so it should not be eaten when water is scarce. Because of its laxative effect, you should only eat small amounts of seaweed at a time. The main types of seaweed available to the survivor in the open seas are listed below:

ROCKWEED: a coarse, dark-green seaweed with large air bladders. It has no food value in itself, but in it and under it you may find small crabs and fish.

KELP: has a short cylindrical stem and thin, wavy olive-green or brown fronds. Found in the Atlantic and Pacific Oceans, usually on submerged ledges and rocky bottoms. You should boil it before eating.

IRISH MOSS: has a tough, elastic and leathery texture, though it becomes crisp and shrunken when dried. Boil before eating.

DULSE: has a short stem which broadens into a thin, broad, fan-shaped expanse that is dark red and has several round-tipped lobes. It is usually attached to rocks or coarser seaweed. It is found on both sides of the Atlantic and in the Mediterranean. It has a leathery texture and is sweet to the taste.

LAVER: purplish to red with a satiny sheen. It is found in abundance in the Atlantic and Pacific Oceans.

Plankton consists of minute plants and animals that drift around or swim weakly in the oceans. They can be caught by dragging a net through the water. Their taste depends largely upon the types of organisms that predominate in the area: if the population is mainly fish larvae the plankton will taste like fish; and like crab or shellfish if the population is mainly crab or shellfish larvae.

Plankton contains considerable quantities of protein, carbohydrates and fats. However, because it also contains chiton (molluscs with shells) and cellulose, plankton cannot be immediately digested in large quantities. If you are living solely on plankton, therefore, you must eat small quantities at first. In addition, you should ensure you have an adequate supply of drinking water: digesting plankton will use up your body fluids.

Each plankton catch should be thoroughly checked before you eat it: remove all jellyfish tentacles (be careful not to get stung), discard the gelatinous plankton (their tissues are predominantly composed of saltwater), and check for species that are spiny. If the catch contains large amounts of spiny plankton, you can dry or crush it before eating.

IMPROVISED RAFTS

Rafts are handy for river travel, and they can also used for short sea journeys, such as travelling between islands in the tropics. Some small rafts, such as the brush raft (see below), can be constructed very easily, though they are not recommended for long-range travel.

Rafts will not capsize easily if they are made properly, but you must remember the following points before setting out on your journey:

☐ Test your raft soundly in safe water before setting out.

☐ Tie all equipment securely to the raft or to the safety line. All personnel should also have a bowline (see Ropes and Knots Chapter) around the waist and secured to the raft.

When travelling by raft you must be aware of the following dangers:

SHARKS: sharks have been dealt with in detail above, but take care not to attract them to your raft by throwing waste into the water when they are around.

SALTWATER CROCODILE: found throughout Southeast Asia, it is a well known maneater. It lives in saltwater or brackish water and it is common near river mouths and along coasts, though it has been sited up to 64km (40 miles) out to sea. Normally around 4.5m (15ft long), it can be vicious and aggressive, especially females with nests. This animal can pose a great danger to you and your raft: keep a watch for it when you are landing or fishing.

CORAL: normally found in warm waters, along the shores of islands and mainlands. Try to avoid it: it can destroy a raft and injure its occupants. If this is not possible, try to steer the raft around it.

SHIPS: your raft is a very small object in a very large sea. Ships may have difficulty seeing it, especially in inclement weather or at night. Take care that you are not hit by a ship: have your signalling equipment ready.

Brush raft (Diagram 113) This flotation device will support around 105kg (250lb) if made properly. You will need ponchos, fresh green brush, two small saplings and a rope. First, tie off the neck of each poncho with the neck drawstring. Attach the ropes at the corners and sides of each poncho and ensure they are long enough to tie with the rope on the opposite corner or side (A).

Second, spread the poncho on the ground and pile fresh brush onto it until the stack is about 46cm (18in) high. Pull the poncho neck drawstring up through the centre of the stack. Make an X-frame of the two saplings and place on top of the brush stack. Tie the X-frame securely in place with the poncho neck drawstring. Pile another 46cm (18in) of brush on top of this.

Third, pull the poncho sides up around the brush and tie the ropes diagonally from corner to corner and from side to side (B). Spread the second poncho, tied-off hood up, next to the brush bundle.

Fourth, roll the brush handle onto the centre of the second poncho so that the tied side is down (C). Tie the second poncho around the brush handle in the

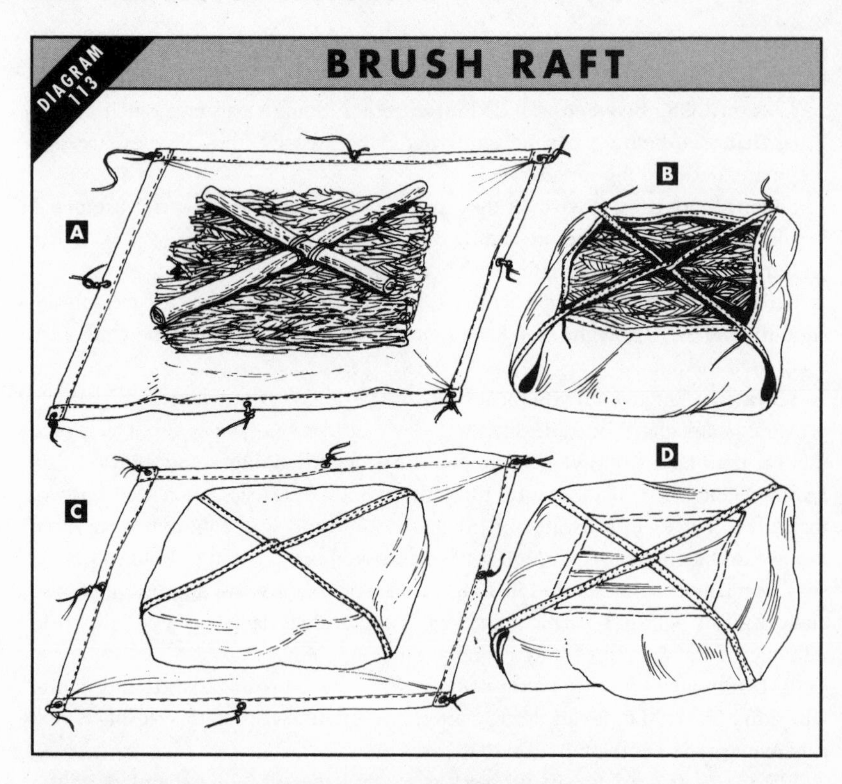
DIAGRAM 113
BRUSH RAFT
A
B
C
D

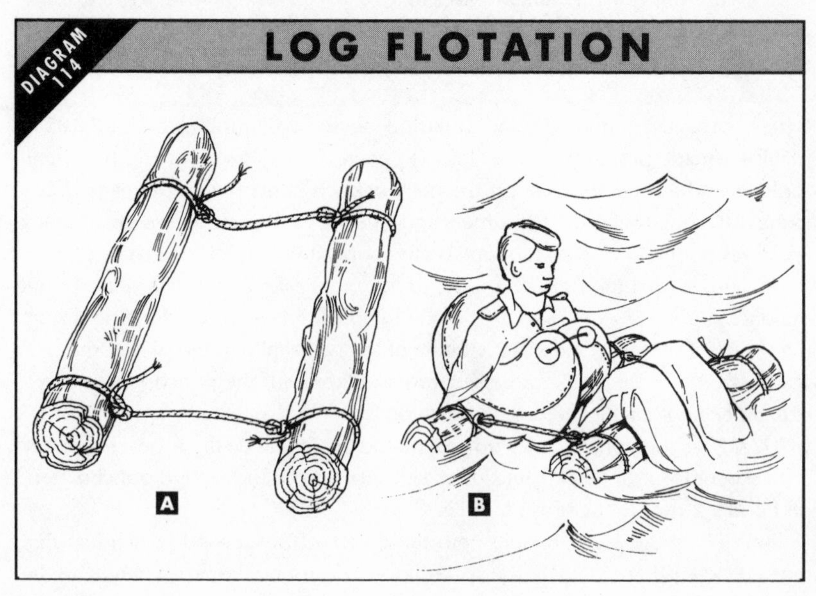
DIAGRAM 114
LOG FLOTATION
A
B

same way to tie the first poncho around the brush (D).

Log flotation (Diagram 114) A flotation device for the single survivor. Fashion it out of two logs of light wood. Place the logs together about 0.6m (2ft) apart and tie them together (A). You will be able to float on them (B).

Vegetation raft (Diagram 115) This raft is made out of small vegetation that will float. Place the plants, such as water hyacinth or cattail, in material or clothing to form a raft for equipment or personnel (it will not hold heavy weights, though).

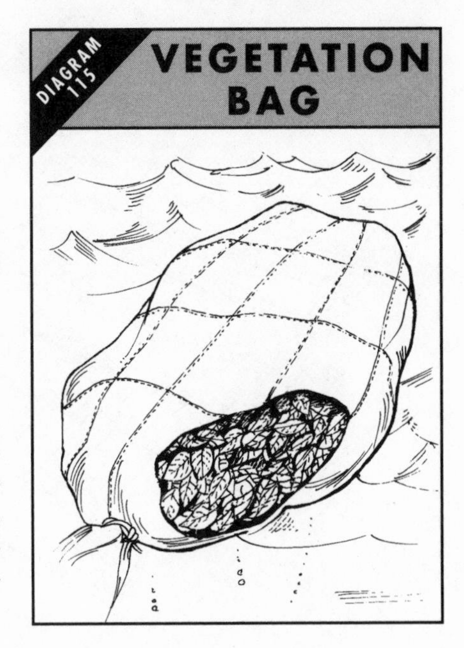

Bull boat (Diagram 116) This is a shallow-draft skin boat shaped like a bath. Construct an oval frame, like a canoe, from willow or other pliable wood. Cover this framework with skins or other waterproof material.

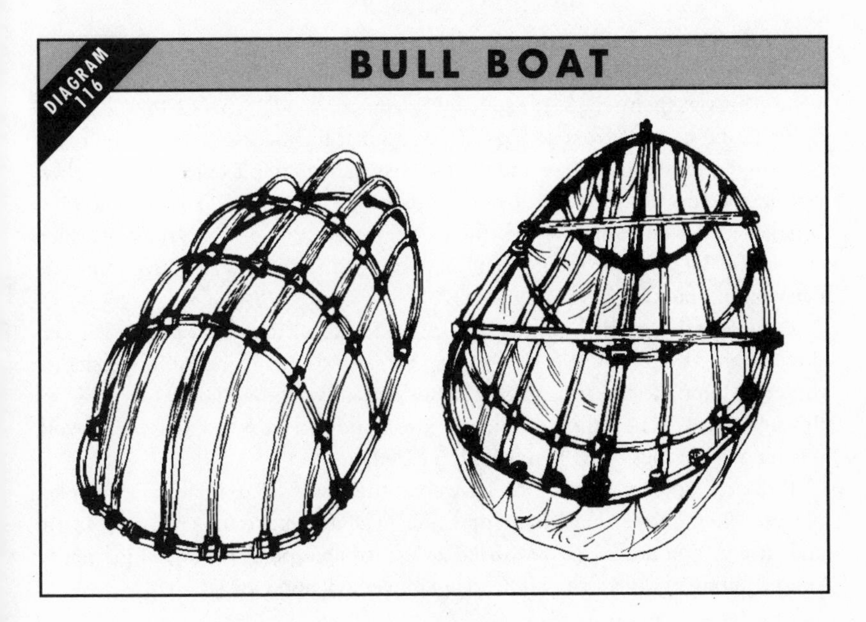

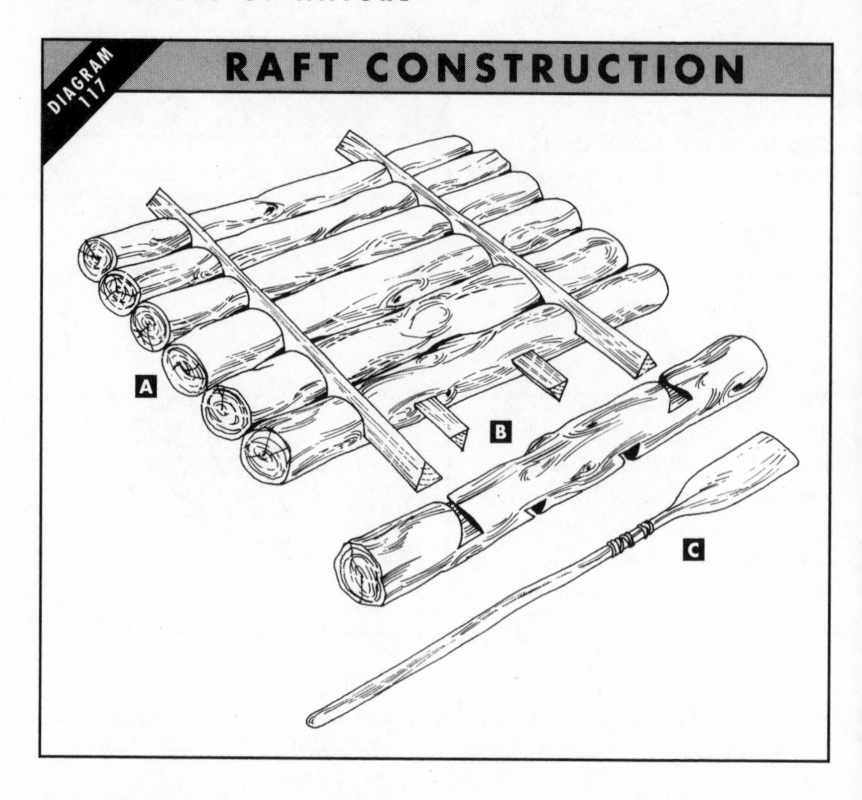

Log raft (Diagram 117) To make this raft you only need logs, an axe and a sheath knife. To carry three people it should be 3.6m (12ft) long and 1.8m (6ft) wide. The logs themselves should have a diameter of 30-35cm (12-14in).

Build the raft on two skid logs placed so that they slope down to the bank. Smooth the logs with an axe and cut two sets of slightly offset inverted notches, one in the top and bottom of both ends of each log (A). Make the notches broader at the base than at the outer edge of the log. A three-sided wooden crosspiece 30cm (12in) longer than the total width of the raft is driven through each end of the four sets of notches (B).

Complete the notches on the tops of all the logs, then turn them over and drive a three-sided crosspiece through both sets of notches on the underside of the raft. Complete the top set of notches and drive the additional crosspieces through them. When the crosspieces are immersed in water they will swell, resulting in the logs being tightly bound together.

If the crosspieces fit too loosely, wedge them with thin, board-like wooden pieces. When immersed in water they will swell and make the crosspieces tight and strong. You are advised to make a deck of light poles on top of the raft to keep equipment dry and a paddle to aid movement and navigation (C).

Lashed log raft (Diagram 118) If you have rope available you can construct a simple log raft. Use pressure bars lashed securely at each end to hold the logs together. The construction of this raft really requires two or more people because the gripper bars are under tension. If you are on your own exercise great care when making this raft.

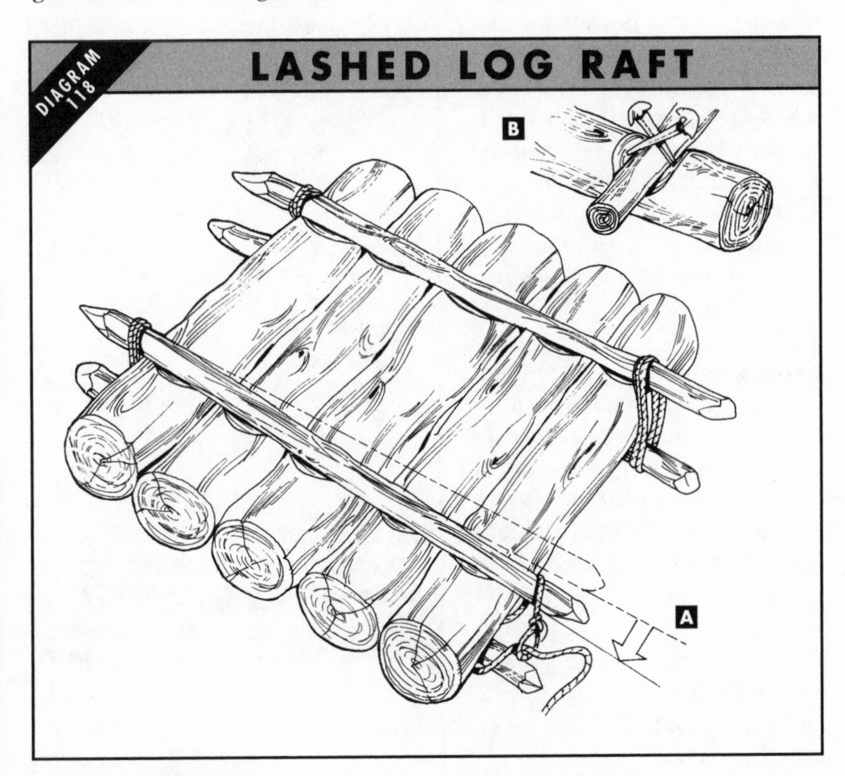

BRITISH SBS TIPS

LASHED LOG RAFT (DIAGRAM 118)

The lashed log raft is relatively easy for two or more people to make and sturdy when completed. Here's how to construct it.

- Place two thickish stakes on the ground and lay logs over them.
- Place the other stakes on top.
- Tie each pair of stakes firmly together on one side.
- With a helper tie the other ends together so that the logs are gripped between them (A), or use hardwood spikes to secure logs (B).
- Notch the ends of the gripper bars to stop the ropes from slipping.

Sea anchor (Diagram 119) A sea anchor can act either as a drag by slowing down the rate of travel, or as a means of travelling with the current. When the anchor is closed (A), it forms a pocket for the current to strike and will propel the raft in the direction of the current. When the apex of the anchor is opened (B), it will act as a drag and the raft will stay in the general area. Adjust your anchor so that when the raft is on the crest of a wave, the sea anchor is in the trough of a wave (Diagram 120).

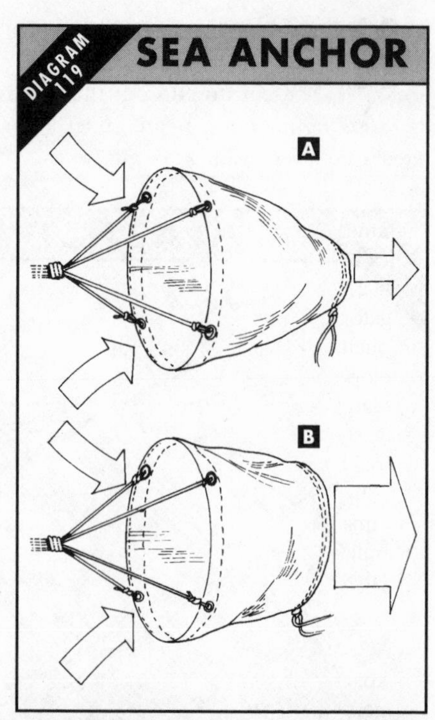

Rafting ashore If you are in a one-man raft, making a shore landing will usually present no difficulty. However, if the surf is strong you could run the risk of capsizing. In this situation you should sail around and look for a sandy, sloping beach where the surf is gentle.

Do not land when the sun is low and shining into your eyes, and avoid coral reefs and rocky cliffs (reefs do not occur near the mouths of freshwater streams). Do not make a landing at night: you will not be able to see any dangers until it is too late. Deploy the sea anchor to help prevent the raft capsizing, though don't deploy it when travelling through coral. If the raft turns over in the surf try to grab hold.

Landings on sea ice should only be attempted on large, stable floes. Stay away from icebergs, small floes and disintegrating floes.

INDEX

T

UV

W

XYZ